Business Operations Unlocked

Jason M. Brown

Business Operations Unlocked

7 Steps to Slash Costs and Streamline Processes with Technology

https://www.jasonunlocked.com

Jason M. Brown

Published by Jason Unlocked Publishing

30773 Milford Rd #171, New Hudson, MI, 48165, USA

customerservice@jasonunlocked.com

www.jasonunlocked.com

Softcover ISBN: 979-8-9934866-0-4

eBook ISBN: 979-8-9934866-1-1

Library of Congress Control Number: 2026905361

First Edition: February 2026

Printed in the United States of America

This book is designed to provide accurate and authoritative information on the subject matter covered. It is sold with the understanding that the publisher and author are not engaged in rendering legal, financial, or other professional advice. If legal or other expert assistance is required, the services of a competent professional should be sought. Nothing in this book should be considered as a guarantee of success, results vary based on implementation and business context.

Disclaimer: References to third-party websites, tools, companies, or materials (e.g., UiPath, Lucidchart, jasonunlocked.com, MIT reports, McKinsey, Deloitte, or other cited sources) are for educational and informational purposes only and do not constitute endorsements, affiliations, or guarantees of accuracy by Jason Unlocked Publishing or the author. The publisher and author are not responsible for the content, availability, or updates of external resources. Trademarks and registered trademarks mentioned are the property of their respective owners.

The examples and anecdotes in this book are based on the author's general experiences and are not intended to represent any specific company, client, or proprietary data. Any resemblance to actual entities is coincidental and not meant to disclose confidential information. Some data and information have been changed to maintain confidentiality.

For permissions or inquiries, please contact: permissions@jasonunlocked.com

Cataloging Data: "Brown, Jason M. Business Operations Unlocked: 7 Steps to Slash Costs and Streamline Processes with Technology / Jason M. Brown. -- First Edition. – New Hudson: Jason Unlocked Publishing, 2026. -- ISBN 979-8-9934866-0-4"

Table of Contents

Table of Figures

Introduction

Introduction

Imagine staring at a shiny new AI tool promising to revolutionize your business by slashing costs, supercharging efficiency, and catapulting you ahead of the competition. Six months later? It's gathering digital dust, your team is frustrated, and your budget has taken a big hit.

Sound familiar? You're not alone. A 2025 MIT report[1] reveals that a staggering 95% of generative AI pilots crash and burn, failing to deliver any real value. Billions get wasted, and opportunities squandered. It's enough to make any leader question whether tech is a savior or just another shiny trap.

Now imagine yourself with all the tools and knowledge needed to navigate these new uncharted AI waters. You have a proven system, teams in place, and technology that is constantly adapting to help your people achieve new levels of efficiency. Because you have all of this in place, you're able to focus on the things that really matter like customer satisfaction, company strategy, and new industry trends. This book was created to set you on the path to that future vision of yourself.

There is a way to implement AI and other technologies not as a risky gamble, but as reliable tools for meaningful growth without overwhelming your budget or daily operations. The processes, tips, and tools discussed in this book are based on my 15+ years of experience delivering products and leveraging technology from bare bones Excel macros to larger ERP integrations, to serverless AI enabled software applications.

---

[1] (Estrada, 2025)

Through years of learning, experimenting, and adapting, I've refined a practical process to improve business operations using technology. Most importantly, I have used this process to generate results that have exceeded expectations consistently. If you ask me how I was able to achieve success throughout my career, this process has a lot to do with it. It's not theoretical, it's based on real world results.

I'm Jason Brown, an engineer, entrepreneur, and executive with over 15 years of experience turning customer problems into profitable solutions. From designing helicopter armor saving lives to launching innovative high-voltage vehicle batteries into new emerging markets, I have led teams through the storm building unbreakable cultures, slashing waste, and driving revenue higher through design of new products and efficient go to market strategies.

Using simple macros in Excel just like the ones discussed in this book, I was able to cut through tens of thousands of lines of data to find gold that not even the CRM system could show (more on this later). At the same time, these basic macros improved forecast accuracy to 90%, helped us consistently grow market share over 10% per year, and fostered a team culture that attracted top talent like a magnet.

No magic, no massive investments required. Just practical, foundational steps that can grow with you. This book isn't theory; it's a straightforward guide based on what I've seen work, ready for you to adapt and apply. It describes my proprietary 7-step arc to master process, cost, and tech management for lean, unstoppable operations.

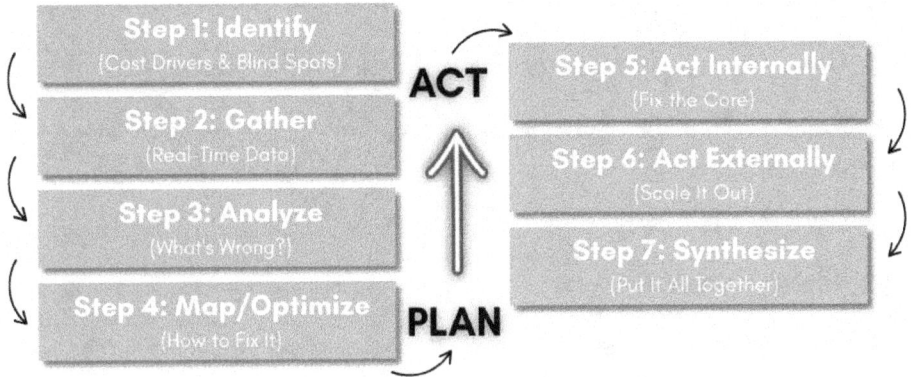

Figure 1: 7 Step Strategy

We'll kick off the process with a deep dive into the cost drivers and blind spots that are hiding your path to progress (Step 1), then arm you with affordable data tools like simple sensors and spreadsheets that gather data in real time (Step 2), and decode the patterns that unlock breakthroughs (Step 3).

Next, we will charge into action mapping and optimizing your workflows with easy flowcharts before integrating AI-driven insights (Step 4), then move on to streamlining your internal ops for quick, game-changing wins (Step 5), and extend outward to suppliers, sales, and markets (Step 6). Finally, we weave it all into a dynamic living strategy (Step 7), one that empowers your team, tracks KPIs, and adapts on the fly.

I've also included some additional topics to refine your approach further such as ethics, sustainability, and pitfalls in the Additional Reading section after Step 7 when you're ready to go further.

To get you up and running quickly, I've also included a ready-to-use Excel Workbook and Supplier Management tool complete with AI prompts to take you behind the scenes of how I put together macros that automate audits and flag risks in seconds.

Quick note: If you haven't received the free extra resources head to:

https://www.jasonunlocked.com/pages/bou-resources

By the end of this book, you'll have the perfect process to leverage technology instead of worrying about it. Let's dive in and unlock your business operations.

Who This Book Is For

This book is crafted for senior executives and business owners in mid-sized firms who feel overwhelmed by operational chaos and fed up with complicated technology initiatives. It takes beginners (those new to ops optimization or tech integration) from foundational awareness to novice-level execution, delivering quick wins like 10% cost reductions through simple tools (e.g., Excel macros and sensors).

For intermediate users it scales advanced tactics such as automation, optimized IT structure, prediction models, and more that can unlock 20%+ efficiencies and gains. Expect a step-by-step transformation: You will map costs, gather data, analyze patterns, and synthesize strategies, all without millions in consulting fees.

This book is not for: Advanced data scientists seeking algorithms. It doesn't contain pages of code and theory. It is streamlined with actionable information based on real world experience for business leaders ready to start simple, build momentum, and achieve tangible results in months, not years. And, because there is only so much that can fit in a book, you can always reach out for more specific help at:

https://www.jasonunlocked.com/pages/consulting

Also, if you're looking for some additional help and want to see these techniques in action, you can head over to my YouTube channel at:

https://www.youtube.com/@JasonBUnlocked

Unlocking Operations: A Holistic Path to Success

Before we get started, I want to explain how I think about business operations because it's probably not exactly what you're expecting. In this book, "unlocking operations" goes beyond the traditional view of shop floors, logistics, or isolated departments. Instead, we will take a comprehensive approach to align any business operation to technology, data gathering, analytics, and strategic synthesis for success.

With that in mind, I want you to expand your thought past your own areas of responsibility. To get the most out of this book, it's best if you put your CEO hat on and think big. You might not have direct responsibility over some of the areas we discuss, but I still want you to try to think about how the business parts fit together. This is crucial because operational problems often span multiple areas of the company. Restricting your vision to just your own area will limit the impact you can have with this process.

Being able to shift your perspective to the bigger picture is extremely valuable to a company, and it's gotten me more than one promotion throughout my career. The CEO and other executives will thank you for looking into gaps in other areas of the business. It never hurts to respectfully look across boundaries to see how teams can help one another perform better. When done correctly, it shows you have foresight, drive, and desire to take on larger areas of the company positioning you for promotions.

When you start to take this perspective, you'll understand that at this level, operations mean everything needed to generate the return on assets. It's a holistic view of the company, not focused on one or two areas. When you shift your view this way, it opens new ways of thinking that get around the typical red-tape and bureaucracy that we often encounter.

For example, when analyzing logistics start with high-level questions such as what is the best way to deliver product to the customer? Framing the question at a high level without putting restrictions on it enables the best possible strategy to come forward first regardless of existing limitations. Then, dig into the processes to make it happen. It shifts to a top-down approach that enables clear thinking and new innovative ideas.

If you want more help in this area, I've developed a Strategy Design Kit that walks step by step through this process of top-down strategy design and implementation that turns uncertainty into action and confidence. You can find it here:

https://www.jasonunlocked.com/products/strategy-design-kit

To help us further with this mindset shift, I'll introduce a new Factory Business Model that I've developed that will make it easier to categorize and analyze across boundaries.

One other key point before we dive into the process: I want you to start thinking of technology like leverage or debt. Taking on debt when the underlying business processes are bad will quickly get you into trouble. This is because the debt is funding processes that don't generate positive returns. Eventually bad things happen like layoffs, waste, and worse. The same applies with technology.

Technology is never a bolt-on solution (even though many companies treat it that way). It's the **accelerator**, turning your foundational framework into predictive insights and automations that scale successful processes into big wins. If your foundation isn't solid, adding technology won't help much. In fact, it can make things worse. This is why the majority of initiatives struggle, and it's the reason we start at the beginning before jumping into advanced technology solutions.

You might think Excel macros are too basic but if you can code in Excel, you can begin to code anywhere. And, most people already have access to Excel, use it regularly, and Visual Basic (the code that Excel runs) is somewhat easy to learn and apply. Plus, you already probably have some data in Excel and Excel can pull in data from various sources easily (more on this later). So, we start by identifying the problems, gathering data, analyzing data, fixing the underlying processes, **then** implementing the changes, and iterating our strategy over time.

The result? A business that's not just efficient but unstoppable. Agile enough to pivot amid disruptions, scalable to handle growth without waste, and resilient through ethics / sustainability. With bonuses like the Lean Strategy Workbook and AI prompts, you'll apply this holistically, achieving 10-20% gains across your entire operation. And, because you're putting in the hard work, you won't need to spend millions on consultants. With this book, you will be in control, and you are going to unlock your business operations.

> **Pro Tip:** Download your included resources now and get ready to implement your first macros using AI prompts at:
>
> https://www.jasonunlocked.com/pages/bou-resources

Step 1:

Map Your Cost Drivers

Step 1: Map Your Cost Drivers

Now let's dive into the details of the process. If you are like many executives (including me), on any given day you're typically knee-deep in your business's daily grind while upper management is pushing targets higher and ownership is demanding more profits to justify investment in your future plans.

You're chasing efficiencies and gains like a detective on a cold case, but every lead and initiative seems to fizzle out. Costs creep up, processes snag, and you're left wondering, "Where's the leak?"

Enter the timeless wisdom of Peter Drucker: "What gets measured gets managed."

Without crystal-clear visibility into operational costs, it's easy to miss opportunities for improvement. That's why we launch this playbook with Step 1: Mapping Your Cost Drivers. This is the beginning step to a full understanding of how and where to deploy your fixes and improvements that come later. It's the start of your roadmap to uncovering those hidden inefficiencies that are silently sabotaging your bottom line.

To maximize the success of our efforts, I want to introduce a business model concept that I have developed and used regularly to generate customer focused business strategies. It will help us categorize all our efforts in an easy to remember framework. This will be important as we layer on more complexity later, so that's why I'm introducing it now in the first step.

The Factory Business Model

Regardless of if your business is an actual factory or not, imagine your operation as a humming factory: Inputs (raw materials or data pouring in), Core Processes (where the magic happens, adding your unique value), Outputs (the products or services flying out), Returns (the cash, relationships, and customer feedback), and Infrastructure (the backbone like tools, tech, and teams making it all possible).

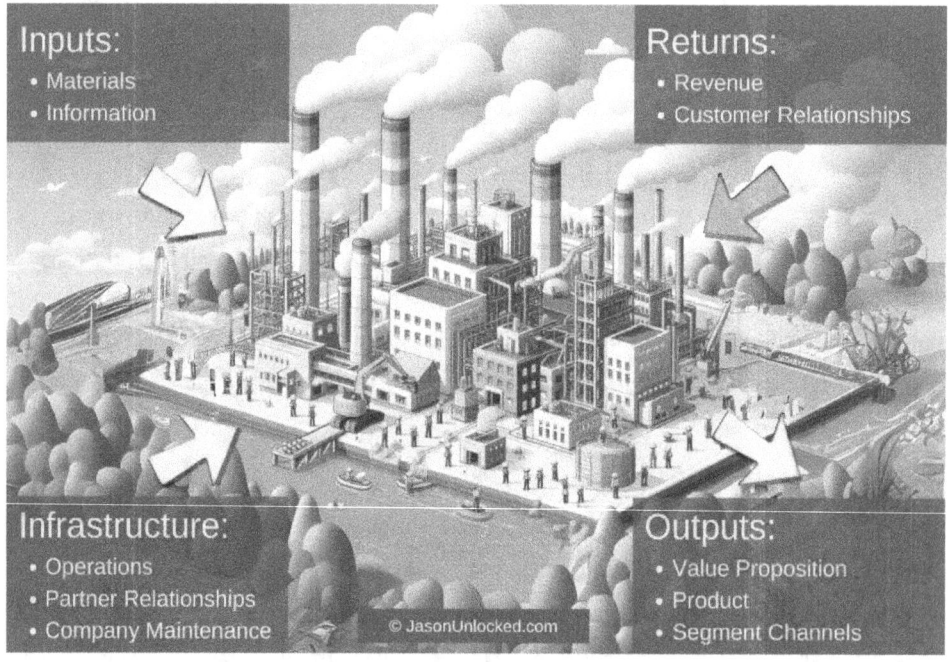

Figure 2: The Factory Business Model

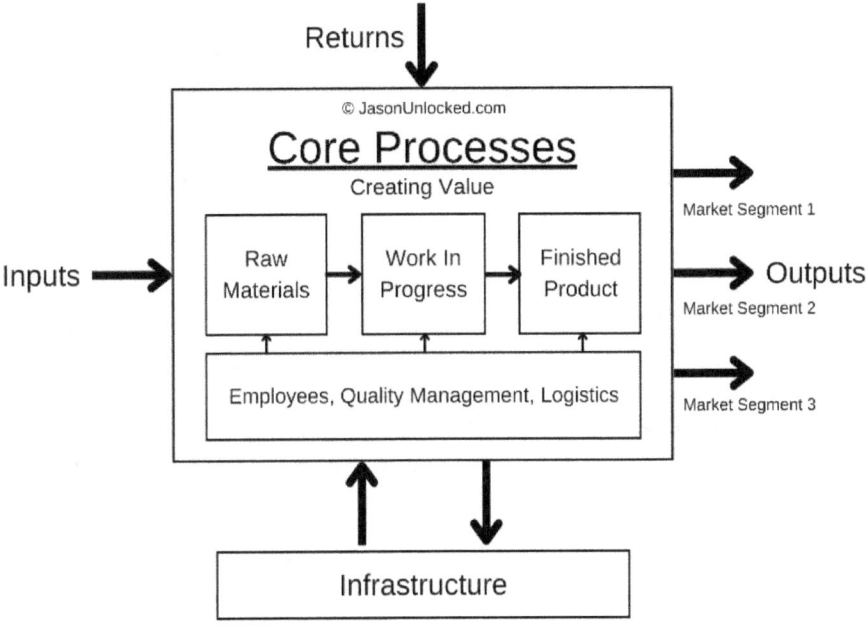

Figure 3: Factory Core Processes

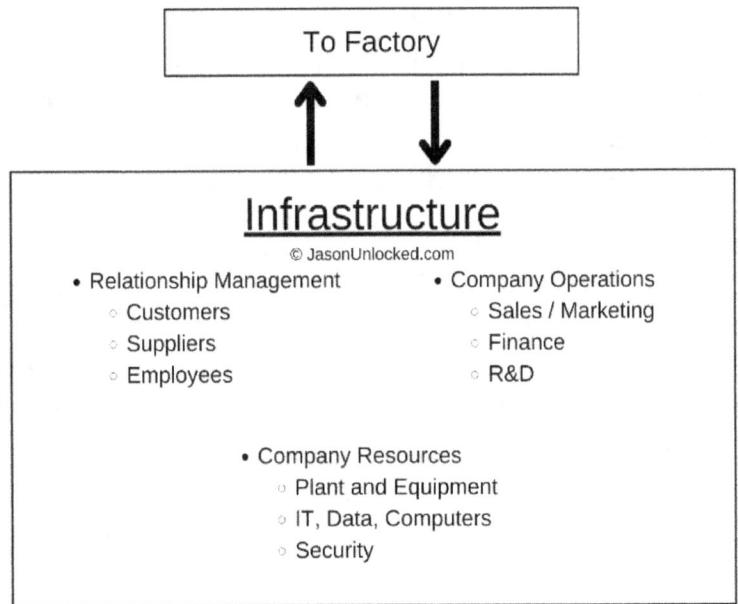

Figure 4: Factory Infrastructure

The most classic example is an actual factory building products that people love to use. But you can take it a step further. Maybe you're a coach or adviser, in which case your process is driving a transformation in a specific person or company. Your input is the person / company before your process, and the output is the person / company after your unique process is applied.

Alternatively, maybe you're an author writing a book. In this case your inputs are the thoughts, ideas, and data and the output is the finished book after your unique storytelling process.

Here is another example: Imagine your consulting firm as a factory. Inputs: Client briefs and research data. Core Processes: Analyzing problems and brainstorming solutions. Outputs: Delivered reports or strategies. Returns: Client feedback, referrals, and payments. Infrastructure: Tools like Zoom for meetings and your team's expertise.

This model can be used anywhere and I find it extremely useful because it puts 100% focus on what matters most: Inputs, Core Processes, Outputs, Returns, Infrastructure. Speaking of what matters most, **Always remember to focus on the core processes that add the unique value to your outputs.** Core processes need to be prioritized while everything else needs to be scrutinized.

Why? Customers pay more for unique value they can't get anywhere else. For example, Apple takes low cost hardware components and adds in their unique style, design, and operating system to create a one of a kind product that people love. Microsoft takes code and knowledge and translates it into an easy to use operating system used all over the world. If you think about any product you use, it can be defined this way. This is a critical step, shift your mindset to put the key value add core processes front and center. Everything else is supposed to support these core value add processes.

For example, infrastructure supports the core processes. The core processes don't exist to justify infrastructure spend. This is important because as we begin to map costs, gather data, and the rest of the steps in this book, it will help us categorize, prioritize, and make better decisions about how to take action later.

By separating these areas early, you avoid the common pitfall of blending them, which muddies insights and slows decisions. No cross-contamination means cleaner visibility from the start, setting you up for the wins we'll build in later steps.

Now let's begin the cost mapping process. This will set the stage for the data gathering activities that come next in step 2.

List Key Cost Drivers

In this book we're going beyond the traditional accounting term "Cost Driver" (which focuses on allocating indirect costs via activities). In this step we will take a broader, more practical view of your entire operation for the expenses that have a big impact on the bottom line. The main point is to identify these drivers early on so that we have clear targets to gather data on, analyze, and optimize later.

Start with what you know. For your core processes, work upstream to the inputs for those processes. If you haven't gotten to the raw materials coming into your factory, go further. Build a larger picture of the company's core processes and inputs so you can find all the significant cost drivers. Start a basic spreadsheet with the Factory categories to jot down data. Right now, we're just trying to identify and prioritize the potential problem areas. Later, we'll track, analyze, and refine them (We'll automate this affordably with tools like UiPath in Step 2, pulling daily feeds from spots like Yahoo Finance straight into Excel for effortless dashboards.)

Assess Impact (High/Medium/Low)

Now, categorize and prioritize the cost drivers. High-impact drivers? Costly and mission-critical (think supply chain disruptions that halt production). Medium? Noteworthy but not catastrophic. Low? Cheap or peripheral. Be sure to go beyond just the data, rally your team for interviews, time key tasks, and run "what if" scenarios (e.g., "What if our suppliers raise prices 20% next year?"). This builds a rock-solid grasp of how drivers interplay with your core processes priming you for the data dives and strategies in Steps 4-7.

Complete the Map

Tie it all together: Link each driver to its factory zone and core process. At this point you don't have to have a perfect data set or factory model. This step is meant to give us a starting point that we'll refine and pivot as needed later after we start getting data and optimizing processes.

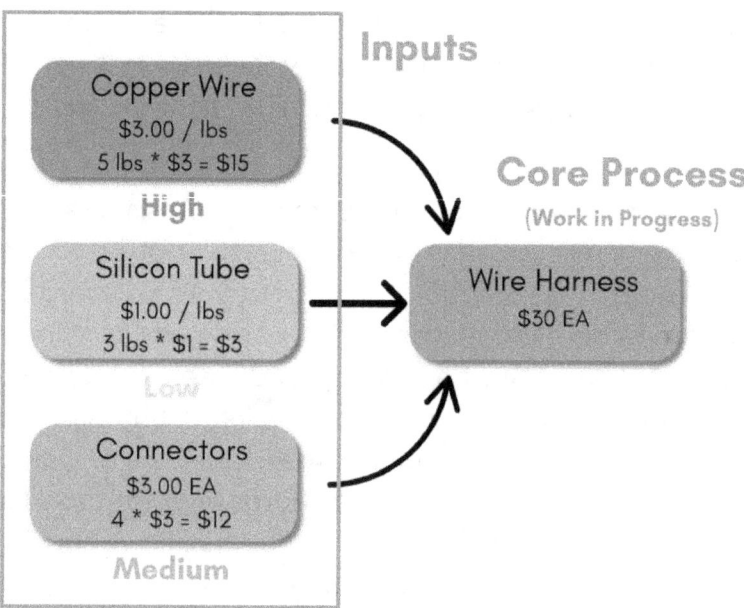

Figure 5: Example 1 Mapping Cost Drivers

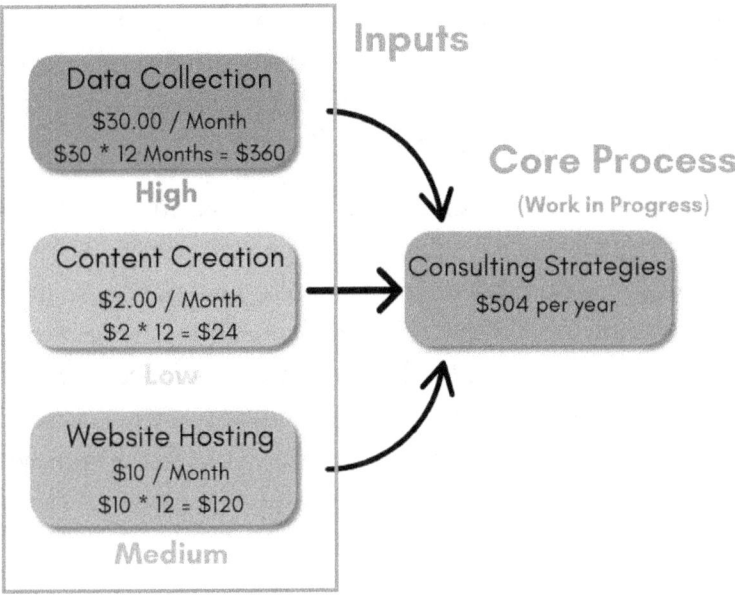

Figure 6: Example 2 Mapping Cost Drivers

Pro Tip: For your top drivers, sketch a quick visual flowchart (pen and paper works wonders, or dry erase board). Then, digitize in spreadsheets for dynamic updates and cost calculations. Tools like Lucidchart shine here for pro-level diagrams; more on this in Step 4.

This simple first step can produce meaningful results fast. At one point in my career it helped me manage a 20% inventory overage on low priority parts draining cash like a leaky faucet. Because I had analyzed the drivers and core process, it was immediately clear that we had a problem. We targeted it with a refined inventory management process that reduced buying frequency and quantity freeing up cash for other uses.

Plus, curbing excessive inventory trims waste and emissions syncing with green tech initiatives and sustainability. Every untied dollar from excess inventory can be redeployed to innovation, making your business leaner, greener, and primed for profit.

When mapping cost drivers, don't forget other areas like labor. The time sunk into activities isn't just a line item; it's opportunity cost. What could your people be doing to make real progress in the company if they didn't have to spend time on less important work? For example, reducing data entry for sales reps frees them up to identify and close more opportunities. A process that can do that will lower costs *and* unlock more revenue. That's the win-win lens of the Factory Model: Pinpointing drivers that ripple across your whole operation.

As we move forward, be sure to focus on the cost drivers with the highest impact first. It's ok to neglect some of the less important cost drivers at this point. Also, I recommend starting small targeting one or two critical cost drivers and core processes to understand them as well as possible. Then, you can expand this process later once you have experience and results to go forward with. Run through this 7-step process with just a few targets first to get a feel for how it works for your situation before expanding to other areas. This will set your foundation faster and get you ready to scale.

With your cost drivers mapped via the Factory Model (sketches or spreadsheets in hand), you're armed with visibility that starts to cut through the fog. Next up in Step 2, we deploy low-cost IoT (Internet of Things) sensors and devices that connect physical objects to the digital world for real-time data. This will eventually enable automation to infuse real-time decision making into your plan to create systems that lead to insight and action.

Step 2:

Deploy Data-Gathering Technology

Step 2: Deploy Data-Gathering Technology

Welcome to Step 2: Deploy Data-Gathering Technology. In this step, we'll begin to build the infrastructure that enables accurate information to flow freely and in real time. Three key themes you'll notice in this book:

1) Start small and build up over time: Set up one project, prove it out, then scale it.
2) Start simple before scaling with complex solutions.
3) Cost effective from the ground up: Use inexpensive and targeted components at the edge, stay away from complex expensive solutions at first.

With Step 1, we started to identify the problem areas of the company and prioritize them. But, even a very detailed analysis of these problems is not good enough in the long run. We need a rock-solid foundation, so in this step we're going to apply hardware and software to ensure we are collecting accurate data in the most efficient manner possible. This is how you create that foundation that clears the path to scale later. Remember, Step 1: Identify, Step 2: Gather, and Step 3: Analyze.

Keeping this in mind, let's dive into an example with a widely available cost-effective tool that your team is probably already using, Excel.

Start with Excel Macros for Internal Data

Excel can be a powerful tool, but most users overlook its real value. Excel has a set of Developer Tools enabled via the options settings of the program. Once enabled, these developer tools allow a user to create macros that run with the click of a button.

If you're not familiar with macros, you can think of them as a way to automate the processes that you already do within Excel.

But how do you get access to all the data in the organization? It turns out that most data programs (including Excel) can output data in what is known as a comma separated value (CSV) file. The reason this is important is to ensure that information can be transferred between the two programs without losing any of the content and context. It's like when two people meet and one speaks German while the other speaks French. They both also happen to speak English, so while it's not their preferred language, they switch to English to have a conversation. Using CSV files is the same thing but between software programs.

Accounting software, CRM systems, ERP Systems, and more typically all have an option to export the reports and databases that house your critical business information into a shareable CSV format.

Excel can import and incorporate these files and their data with relative ease. This allows you to do powerful analysis, like exporting your sales data by month from your CRM or accounting system and then importing it into Excel for analysis. If you have other data, like marketing spend or logistics performance, you can add that alongside the sales data to build out a bigger picture of company performance. Similarly, you could pull in accounting data such as past due customers or quality information on suppliers. **This allows you to start building out your data collection systems to match the core processes and drivers you found in Step 1.** The key benefit is being able to easily pull data from multiple sources and that's where Excel shines.

Now, you may be thinking that something like Excel is too simple and obvious. It can't possibly make a significant difference in processes. There must be a more sophisticated system out there that is better, right?

More sophisticated, almost certainly. Better? That's questionable and here's why, from a real-world example that I see often.

How do you know that your software systems (like CRM) are reporting data correctly? Most people assume that the data coming from their software reports are perfect, but here's the truth: They're not.

This can have drastic consequences for your business. For example, what if your sales team categorizes a customer into two segments? A lot of customers float in between segments, or maybe it was a simple mistake. Does the CRM account for this customer twice in the system, once in each segment? This could show higher sales than actually achieved by double counting the account.

Here's another common problem I see, sales are up but margins are down. What happened? If your CRM system doesn't talk to your supply chain management system, you likely won't know. Or consider this: A customer calls up claiming they didn't receive the full order when your system shows the order was picked and shipped correctly. What do you do? The answer to all of these problems is a simple process to verify information without spending millions on consultants and custom software, and that's where Excel can step in.

Here's the high-level summary of steps to get started:

Figure 7: 3 Step Process for Data Collection

Enable Developer Tools: Visual Basic and CSV Files

The first step is to understand the macro system which uses Visual Basic code. These days, you don't have to be an expert in code to be effective thanks to AI. I've included an example Excel file with AI prompts along with this guide to help you see how AI can do the heavy lifting in creating the macros you need. Use AI as both a teacher, guide, and developer to help you get up to speed fast. Simple coding like we're going to discuss is a great place to start. It should also be able to guide you step by step through enabling Developer Tools in Excel, naming Macros, and getting to the developer environment where the code goes.

To see walkthroughs of how to implement Excel Macros and code, subscribe to my YouTube channel at:

https://www.youtube.com/@JasonBUnlocked

Now, it's true that CSV files make life easier in getting access to data, but that doesn't mean the job is over. After you've enabled tools and found useful sources of data, we must get them ready for analysis because raw data sets often include a lot of extra information that is not needed. Plus, it is likely in a structure that doesn't make sense for what you're trying to accomplish. If you're exporting sales information, for example, it might include the date the report was run, the name of your company and other meta information. Also, the data might be shifted so that sometimes sales are in column B and sometimes they're in column C. Let's talk about how to get your data ready for action.

Import, Clean, and Verify Data

After you've set up developer tools and gathered CSV files, the next step is to get rid of all the extra information. You can do this with simple search macros.

First, have a look at your data and look for patterns like company name showing up consistently in rows that you don't need. You can run loops in the macro to go line by line and search for key terms (like company name) that you want to remove and then delete the row. Excel can do this automatically on thousands of line items in seconds.

One piece of advice: Go one step at a time to gradually transform the data set. For example, write code to remove the customer's name rows first. Then, after you verify it worked successfully, write code on the next task which might be to shift data from column C to column B. This is a safe way to ensure each step is successfully executed and will save a lot of headaches and re-work later. It may seem slow but remember we're building our rock-solid foundation so everything must be verified step by step.

Also, always copy your original data set in case you must go back, for example if the code accidentally removes too much information. The good news is that once you've got this set up correctly, your macro can run in a matter of seconds for every future instance where you want to run the report. You just import the new data and hit the button to clean and organize.

Once you get used to this process, you can expand your analysis to include other programs. As long as they export in a general format like CSV (most do), you can repeat this process to clean and structure and then combine your data sets.

For example, if the accounting or procurement team have reports showing costs per part, now you can add that to your sales report and track changes in margin over time. Clean and organize the data separately first, then combine it after to simplify the code and speed up the process. Once combined, you can filter by account to double-check your sales data and make sure the products all have a healthy margin.

Or consider this example: As a solo software developer or small team. Use Excel macros to track user data of your websites and apps without fancy tools. Export app metrics (e.g., views, downloads, subscriptions) as CSVs from your analytics software. A beginner macro can clean duplicate users with a simple AI prompt: "Write a macro to remove repeated user IDs." This verifies data like unique visitors over time or subscribers per view.

You can also use Excel to track processes as well. For example, you can create macros assigned to two buttons within an Excel file such that anytime a user starts and stops a process (by clicking the corresponding buttons) the current date and time is automatically added to the file in a new row. Additionally, you could have a user input their name and the macros can also put the name next to each entry. In this way, you've got a simple tool that can track the time it takes to complete any task with a click of two buttons. Each row will have a start time, stop time, and name that can be used to build averages and look for variance between users which can then be used to improve processes. This is a perfect example of the process this book teaches. Identify, gather, analyze, improve, act, and iterate.

The possibilities for quick but meaningful gains are at your fingertips once you gain a basic skill set around Excel and Macros.

Address Limitations: Options for Big Data

This is a great way to get started gathering, cleaning, and organizing your data, but what if you need to analyze tens of thousands of lines of data or more? At this point Excel can start to slow down such that your macros are taking minutes to run.

This becomes a problem when you're trying to run several analyses or explore changes and how they affect outcomes. Any time you're iterating (running a lot of analyses), these minutes can turn into hours of time eaten up just running the macros. Not ideal.

When this starts to happen, look at tools like RStudio and Python. These programs are code based, similar to the macros you've developed, but can handle much bigger data sets. And, a program like RStudio has built in data visualization tools and dashboard building capabilities that make it a great way to analyze and see data (more on this in Step 3). It will take more investment to learn the code and program, but the payoffs can be great. In general terms I find that Excel is best for beginners / intermediate with moderate data sets and RStudio is next for intermediate / advanced with very large data sets. Python is more for advanced users because it is all code based so you're going to have to invest a lot of time learning the code if you're not familiar with it.

Remember that AI is always there to help you streamline the code and ensure it works as intended. If you use an IDE like Visual Studio Code, it has AI built into it that will analyze your code directly. But, you can also copy paste code in and out of other AI platforms and then implement it in your Macro or code. If you're not already using AI, start now to get familiar with how it can help you automate your data collection.

In fact, Gartner's 2025 data and analytics predictions highlight that by 2027, 50% of business decisions will be augmented or automated by AI[2]. But here is the thing that trips up most executives. You must be ready with your data (gathered, cleaned, and structured) to get the maximum benefit from these AI enabled processes. AI is not a cure all and it can only deal with what it's given (garbage in, garbage out).

This process is exactly how you get your company ready, by setting up simple systems now with Excel macros, RStudio, and Python and getting a foundational grasp of your data so you can scale with AI later.

Pro Tip: I like to keep my code and AI separate (copy / paste code in and out of an AI platform). I do this to ensure that only the specific changes I want get implemented. It's slower than working directly with AI in the IDE (Vibe Coding), but to me it's more important to ensure high quality than speed. Plus, I learn more when I must analyze the code myself instead of having the AI make the changes. It's better to do this if you're just starting out or want to learn more.

Leverage UiPath for External Data and Automations

Once you get a basic architecture up and running for collecting company data around your core processes and drivers, the next step is to start looking outside the company to make your analysis even more powerful. Data that exists within your company should be easy to get to. But, how can you leverage external data to refine your processes and performance even further?

Well, you should have identified the main cost drivers and processes from Step 1, so now is the time to conduct some research on your cost drivers. If you're looking at things like raw materials (wood, steel,

---

[2] (Keen, 2025)

aluminum, copper, etc.), this information can be found on public websites that list the latest market prices.

You can extend this beyond just cost drivers too. For example, if you sell parts that go on cars, look for information related to auto sales. Find company performance data of a public company like yours or your customers (such as stock price over time or analyst forecasts). This type of data can be great for predicting what your sales might be for the upcoming quarter or year. In step 3 we'll talk about how to find correlations between this data and your company's performance.

Or, if you sell to consumers (like phones, computers, books, etc.) you might want to look for metrics around general consumer health like consumer confidence numbers, debt level / savings level, delinquency rates, and more.

Identifying external data sources is key, but pulling data manually is very time consuming. It's hard enough to manage a company's internal data let alone other public information. Enter technology to save the day. UiPath is a program that allows users to automate processes, similar to macros, except it connects things that are not usually connected (like Excel files and websites).

If you want to regularly (maybe daily) pull the market prices for your raw inputs, UiPath can go to specific websites, pull the tabulated data, and then input it into an Excel file automatically. It can also work with your internal data too, pulling reports from several programs into a central Excel file for analysis.

In this way, your team can be ready to hit the ground running every day. The data will be waiting for them; they just need to run the macros / algorithms. This gets them right into decision making, strategy adjustment, and action.

An easy way to keep track of your new data gathering systems is to use a simple framework like the one that follows:

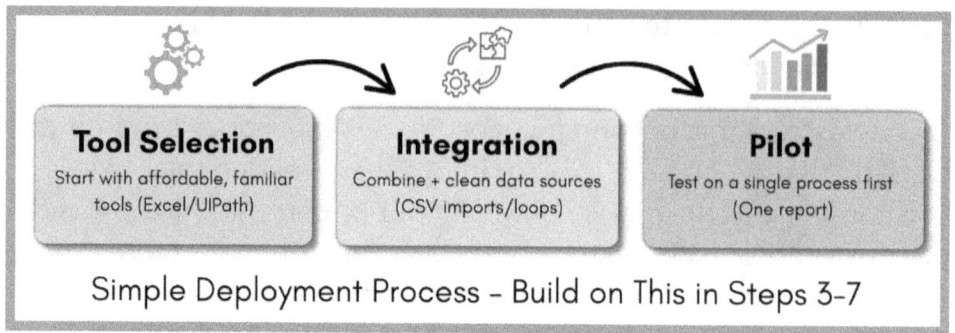

Figure 8: Scaling Data Collection

Add Hardware: Low-Cost Sensors for Physical Insights

Data collection of physical processes is harder, but possible and it doesn't have to be expensive. Most companies overlook this step because they think it adds too much complexity and cost, but here's how you can do it by starting simple:

- **Start with low cost IoT devices:** Microcontrollers such as the Arduino family of devices (~$10-$50 per device) and Raspberry Pi are low cost, flexible, and easy to learn with. Start with a basic project outline by listing goals. Then, map them to devices needed. For example, temperature sensors, light sensors, weight sensors, motors, switches, etc. can all be plugged into an Arduino device that will record data and export to a CSV file for analysis.

- **Build a prototype:** Use breadboards and jumper harnesses to build a proof-of-concept system first. Verify basic functionality such as power draws, switch activation, voltage levels, and read data of the serial monitor to verify accuracy. Begin to troubleshoot and calibrate the components and code.

- **Build in connectivity:** Add data export operations to your devices such as CSV / Json files that can be used with other programs like Excel. This turns the physical measurements into data that can be analyzed later.
- **Finalize and prepare for production:** Once the concept is proven, begin to build out a production level system with custom PCBs, wire harnesses for robustness and safety, and Wi-Fi / Bluetooth for connectivity.

These systems allow you to bridge the digital and physical parts of your business by connecting real time data to your Excel setups via basic interfaces. This turns passive mapping into active alerts such as missing inventory, long build times, sudden price changes, and quality defects.

Remember the Excel file I mentioned earlier that logs process time and username? Well, now you could incorporate physical buttons that someone pushes to start and stop the time logging. This increases the ease of use and therefore accuracy of the data. Instead of opening a file and clicking a button, they push a button. And, buttons can be put almost anywhere to increase the scope of what can be tracked. If you also add in a badge scanner to pull employee information like name, you've got a powerful time tracking system for very low cost.

For more info on how to get started on projects like this, see my smart home brewing project at:

https://www.jasonunlocked.com/pages/ai-powered-smart-brewing-automation-system

In my smart brewery project, we piloted sensors on brewing equipment to monitor processes and power consumption and reduced raw material usage by 10%. Reducing raw material use by 10% not only slashed costs but built resilience against supply spikes, turning sustainability into a competitive edge with real ROI. No fancy overhauls, just some simple edge computing to process data and store in a local database.

Others are catching on to this trend too. Deloitte's 2025 Smart Manufacturing Survey shows similar setups boost efficiency by 10-15% for early adopters, cutting downtime and emissions without big investments.[3]

The key is to start simple, start small, and then scale to bigger wins. For example, in my brewing project I started with one part at a time. First, I verified the sensors and valves had power. Then, I verified that I could turn components on and off. Then, I installed the components and verified functionality. Then, I started to pull data during trial runs. Then, I started to adjust the control to dial in the results. In this way, I systematically stacked wins on top of each other and ended up with a very accurate sophisticated system.

For more inspiration to help you get started with projects like this, Mas Holdings used UiPath to automate 52 processes, driving productivity, efficiency, and cost savings by redirecting resources to higher-value work.[4]

Read more here:

https://www.uipath.com/resources/automation-case-studies/mas-holdings-manufacturing-rpa

By this point you should have some ideas on how you can build out your data gathering system with simple low-cost software and hardware. Doing this work now will put you ahead of the competition. According to Deloitte's 2025 Smart Manufacturing, 92% of manufacturers surveyed said that smart manufacturing will be the main driver for competitiveness over the next three years.

Plus, companies that implemented these types of changes saw on average a 10-20% improvement in production output, a 7-20% improvement in employee productivity, and 10-15% in unlocked

---

[3] (Gaus & Schlotterbeck, 2025)
[4] (UiPath, 2025)

capacity. Furthermore, automated data pulls reduce paper-based tracking and overproduction waste by 10-15%.[5] This systematic approach is the key to ensure that you're not one of the 95% companies that fail with technology initiatives.

With data now structured and flowing in real-time (from Excel automations to sensor feeds), you're primed for breakthroughs. In Step 3, we'll dissect it to unearth trends, benchmarks, and hidden gems, fueling optimizations in Step 4 and beyond.

> **Quick Win:** Set up a UiPath bot for daily market price pulls. Aim for 10% better forecasting in months.

---

[5] (Gaus & Schlotterbeck, 2025)

Step 3:

Turning Data into Decisions

Step 3: Turning Data into Decisions

You've got your cost drivers mapped (Step 1) and data flowing in real-time (Step 2), now comes Step 3: Turning Data into Decisions. This is the step where raw numbers turn into insights that expose flaws, spotlight opportunities, and propel your business forward.

In this step we'll explore patterns to understand why costs rise, where efficiencies might be, and how to anticipate changes. Remember the Factory Model as we go forward. Bucket your analyses by inputs, core processes, outputs, returns, and infrastructure to keep initiatives organized.

From my experience, I've found there are three key steps to go through when analyzing data for a company. The outcome is a solid foundation of understanding which prepares the company to scale later:

1. Verify Information / Truth
2. Analyze Performance
3. Make Predictions / Projections

Master these, and you'll transform data overload into a structured decision-making process. Let's break it down.

Verify Information / Truth

Trust but verify. The first thing to do with your new data is to confirm the existing reports and data that were being used. This is a key first step because if you can't replicate the old reports with your new data, or if the new data doesn't match known truths, you risk making problems worse.

If the new data and processes lead to exact matches of your reports today, then check this box and move on. But, if there are any discrepancies (even minor discrepancies), you must spend time investigating it. Little things can turn into big mistakes and either your new processes are wrong, or your old data is wrong in some way. Both can lead to major problems, so get it right first.

Here's an example: When I first started to implement macros in Excel to evaluate sales data for a company, I found several products that the company was losing money on. Every time they made a sale, the company lost money and no one knew it. Not the accounting team, not the production team, not even the sales team. The root cause was suppliers raising the cost of parts over time, but the system didn't flag it. Because the systems didn't talk to one another, the sales team didn't adjust the sales price and eventually the costs got so high that the margin became negative. After we identified this problem using the Excel Macros and data pulls, we were able to quickly fix the price which resulted in an immediate boost to financial performance.

In another case, I had categorized and totaled sales for all accounts that should be in one market segment, however the numbers weren't matching what the CRM system was showing. This was a clear indication that the CRM report needed attention. Later, I found the companies that were mis-categorized in the system. We updated it, and the team finally started getting credit for their hard work with these accounts.

Here's another example to consider: For a small online shop, verify sales data by cross-checking platform reports (e.g., Shopify or Google Analytics) with Excel imports. Look for mismatches like orders showing as "shipped" but the inventory change doesn't match. If totals don't match, dig in. This process builds trust in data for decisions, like predicting holiday surges. All done with some simple Excel macros.

Even if you don't make physical products this methodology still applies. Similar to inventory checks, Excel can be used to pull in marketing data to track ad spending and conversion rates. Use these basic reports along with an AI program to analyze and improve performance for a 10% better ROI. Trust but verify all of the critical cost drivers and data collected for critical processes.

One key point when comparing your new data to old reports: I always recommend starting with the high-level summaries before diving in to more detailed information. Meaning, look at top-level numbers that should be well known. For example, total sales, total sales by category or segment, total marketing budget, total variable costs / fixed costs. See how your new data rolls up and compares to these high-level summaries.

I start here because it will quickly show if there are discrepancies because everything must roll up perfectly to match. One little discrepancy will lead to a mismatch between new data and old reports. You can very quickly see that either it's perfect, or it needs further attention. If there are issues, it's time to move into the more detailed categories.

Plus, this methodology isn't just for numbers; it can apply to any process. For example, your team thinks it takes 5 minutes to complete a form. Build out a simple system to check (like the simple button Excel program I spoke of earlier) and then compare that to your assumptions. You might find that the numbers are off (for better or worse).

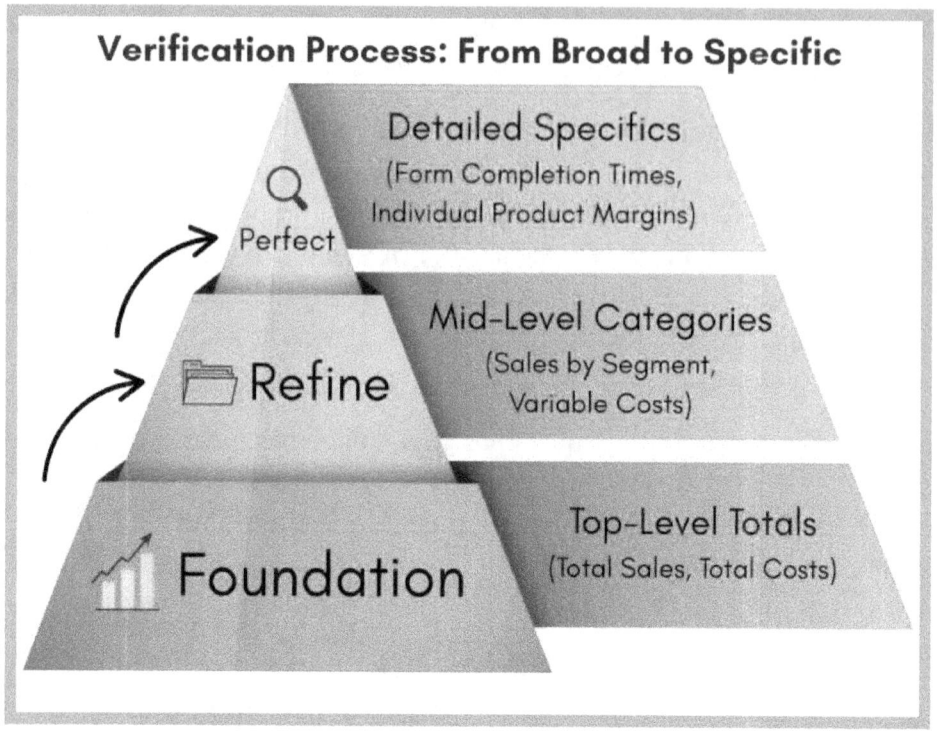

Figure 9: Data Analysis Pyramid

At the end of the day, you need to be able to say: "We know our processes, we know our data, and here's how we know." If you can't do that right now, then spend time on this step first.

Analyze Performance

Once you have a good feel for your numbers and processes, you can dive into analysis. In this book, analysis is used to turn data into insight and action. So, let's talk about how to do that.

One easy way to start analyzing performance in Excel is using the Data Analysis tool. First, within Excel go into the File menu and select Options. Then, select Add-Ins and look for the Manage section. Select Excel Add-ins and hit the "Go..." button. Make sure the Analysis ToolPak is checked and then hit the OK button. This will create a new section in the Data menu tab.

Within the Data menu tab, there will be a section that says "Analyze" and within this you will find the Data Analysis tool. This is the tool that allows you to run regression analysis and a lot more on data sets within Excel.

One great way to use it is the Descriptive Statistics function. Click on the Data Analysis button and then find "Descriptive Statistics". This will open a new window where you can select the data range you want to analyze. Make sure you check the Summary Statistics box and specify an output range (can be one cell for location purposes) and indicate if you have labels in the first cell of the range and how data is grouped. Then, click OK and Excel will output a summary table that shows all meaningful statistical data for your range of data such as Mean, Standard Error, Median, Mode, Standard Deviation, Variance, Range, Sum, Count, and more. It's well worth your time to explore this Data Analysis tool more especially when we come to regression analysis. There are some very powerful tools within Excel that rarely get used.

Using these techniques for my smart brewing project, analyzing sensor data showed that we hit specific gravity targets 30 minutes ahead of time through improved mashing techniques that reduced variability in mash temperature and sugar extraction that led to 20% reduction of electricity used to heat the mash once addressed.

Running a quick check on data with Descriptive Statistics is a great way to spot variability problems and seasonality effects in data. These ups and downs in cost numbers or outputs over time can signal hidden inefficiency or problems, like inconsistent machine run times or fluctuating supplier prices.

Group data by time periods (year to year or month to month) or categories (Product A vs. Product B). Then use Descriptive Statistics to analyze the averages and deviations and flag major differences between time periods or groups. More variability typically needs more attention.

Data analysis is great by itself, but it becomes even more powerful when matched with effective data visualization. Data visualization is a term for the charts, graphs, and illustrations that help us quickly make connections, spot correlations, and tell stories about our data.

There are a lot of books and resources for data visualization that are outside the scope of this book, but I will point out some key lessons when fusing data analysis with data visualizations that I've learned through my career giving presentations to C-Suite audiences and building out business cases to justify investment.

Key Lesson 1: Become Fast and Flexible

One of the major reasons I recommend going simple first is to maximize flexibility and speed. Simple processes that are well understood can be changed in minutes and days vs. weeks and months for complex processes.

Businesses, markets, and consumers can change fast, and you need to be able to adapt just as fast to win. Data visualizations are the best way for the human mind to track progress, find trends, and find correlations.

So, how can we implement effective visualizations that can change and adapt quickly? In keeping with our theme, start simple. Excel has a great set of data viz tools built in, and they almost always are good enough to do the job. This might seem obvious, but here's the catch: Now that you've implemented automated data gathering and filtering with Macros, it will make your basic Excel charts seem like magic. Because, with a few clicks now you can update information in near real time using your macros. As the data updates, the visualizations also update so you can see changes fast. Connecting the data to simple data viz tools in Excel is a great way to get familiar with this process. Because it's well known and simple, you and your team will be able to move fast with it.

For more advanced users, pivot tables are another great tool to use in conjunction with data visualization.

If Excel starts to slow down, or you're looking for more advanced options, RStudio or Python are the next-tier programs that have great data viz tools that are a little more complicated to use, but even more powerful. The key takeaway is to find the right mix of speed, clarity, and refinement in the data visualization.

This is important because you could be looking at dozens of data set visualizations that are changing frequently. And, depending on who you're presenting to, they may want to see things that you haven't thought of yet.

A simple, fast, and flexible process will enable you and your team to bring up new data analysis quickly so that decisions can be made quickly. Don't lose the momentum, become fast and flexible with your data analysis.

Key Lesson 2: Let the Data Do the Talking

One thing I see frequently in meetings (even with high-level executives) is an over-reliance on opinion instead of facts when discussing operational strategy and performance. When you're in a meeting with C-suite executives, you won't change minds and generate support without clear data.

Here are three key pitfalls to avoid when making important presentations with data:

1) A lack of data and facts to support ideas will put the meeting focus on the wrong things such as your abilities, opinions, and credibility instead of putting the focus on the best possible solution to the problems.

2) Remember that these executives are people that may not know you and they don't know the history and all the context of what you're presenting. However, you need them to decide on your plan. This is where a well-defined story with strong data and visualizations shine.

3) If someone contradicts you, you need the facts to back up your position or to pivot quickly. Again, strong data and visualizations are key.

Jump over these hurdles by putting the focus on the problem and the strategy with data.

When coming up with analysis and then strategy, your data should do the talking. What I mean is that anyone should be able to look at your data and analysis and see the direction you're heading.

When done correctly, this approach frees you up to spend your time on the most value-add topics such as past performance, the changes you want to make, the outcomes you expect from the changes, and then the plan or strategy to implement the changes. Stay on point with your message and let the data guide and support the discussion.

Now, the specific analysis and data visualization will be unique to you and your needs. Maybe you want to analyze sales over time. Or, see if there is a Pareto rule with your customers (80% of results come from 20% of customers). Or, maybe you want to track delivery time, or work order completion rate. Remember to go back to your cost drivers, the factory areas they fall into, and then the data gathered. If you follow this process, the type of analysis you need to do should start to become clear.

One powerful data visualization is a chart like the one that follows to identify Pareto rules within the company. The first thing I do when looking into a new company is to look at what is working. Find the key customers or the key activities that are generating most of the results. When you do this, you can start to analyze what makes the company stand out and refine the go-to-market strategy to find other customers and activities like the top 20%. This is great for sales and marketing, but it also applies to other areas too. For your area of responsibility, find the 20% of activities that generate 80% of your yearly results. Then, you can focus your resources on improving these activities to improve performance faster.

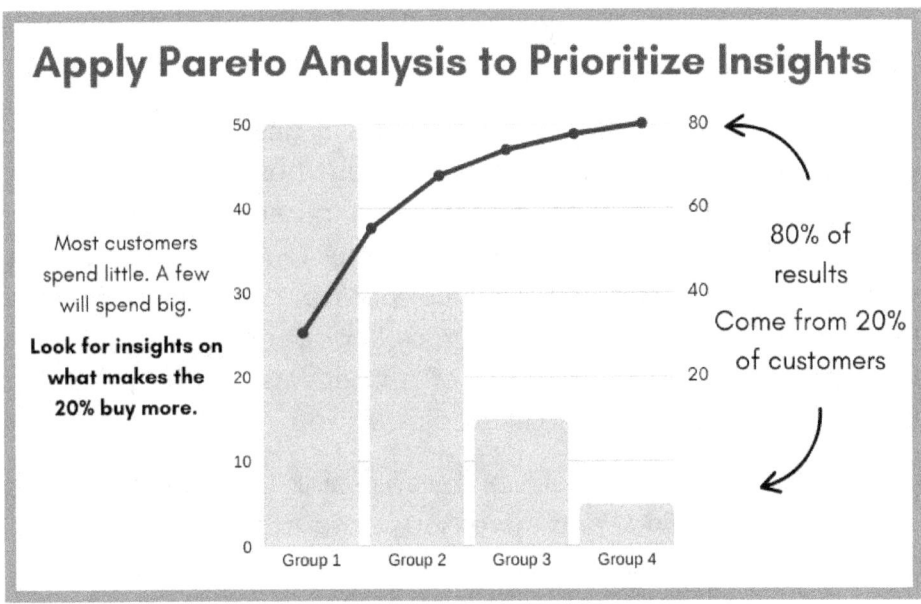

Figure 10: The Pareto Rule

Go Deeper with Analysis: Using Regression Models To take it a step further and spot those hidden correlations try basic linear regression analysis. Start simple: In Excel, use the Data Analysis Toolpak for linear regression. This tool helps identify connections and drivers to transform data from backward to forward looking so you can master all aspects of company performance.

A regression model analyzes how a dependent variable (such as sales) changes when independent input variables are changed. It will tell you how significant each input is to a positive or negative change in the key variable. Put simply, it can indicate things like changes in price have a significant impact on sales. Or, that a higher outside temperature will lead to more sales. It's a powerful tool that can compare several different inputs at once so you can rule out correlations quickly and focus on what matters.

Once you start to find interesting connections, investigate further. This is needed because linear regression can find correlation but not causation. Just because it looks like higher temperatures increase sales doesn't mean it's true (causation). More work is always needed to verify or rule out the findings.

For example, at one point in my career the largest customer segment driving sales was in the mining business. I decided to investigate the price of copper (their main mineral extracted), and I was able to determine through linear regression that our sales were positively correlated to the price of copper. So, I investigated further and found that sales tracked the price of copper but with a six-month delay. Meaning, if the price of copper was trending higher, six months later our sales would likely pick up too.

I was able to verify this correlation by analyzing the time it took for the higher market price and increased demand to filter through the supply chain back to us. The process went something like this:

As the price of copper rose, the mining operations wanted to extract more material faster, so they bought more equipment. The dealers that sold the equipment to the miners had to replace their sold inventory, so they bought more equipment from their manufacturers. Then, the manufacturers increased their builds to have products ready for the dealers. At this point, the manufacturers issued more purchase orders to our company and our sales increased. After I found this correlation and verified it, we could then make better inventory decisions, better production scheduling, and better investment decisions based on the expected demand to come.

The result? We were able to hit 90%+ forecast accuracy and reduce our logistics costs by taking advantage of large order quantities when we needed to.

Plotting out data over time visually can help identify patterns to confirm or rule out the correlation. This was critical for my copper correlation because the graphs were shifted by six months. So, I could visually see that the two variables were correlated, and I needed to shift them to finalize the correlation.

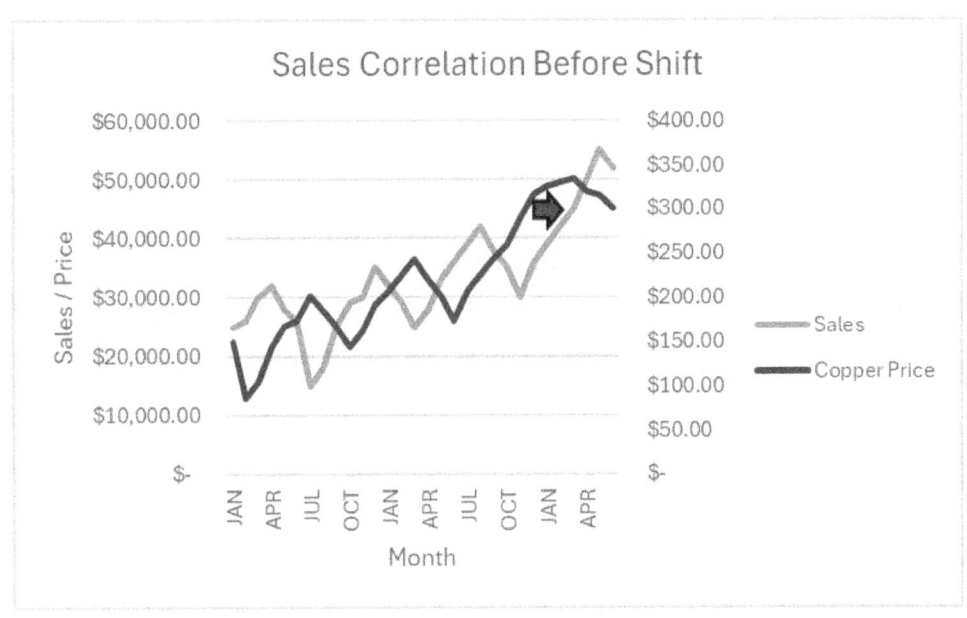

Figure 11: Sales Correlation 1

Figure 12: Sales Correlation 2

In the graphs above, at first there doesn't appear to be a correlation as the two lines seem to be moving opposite one another. This is due to the six-month delay between the price of copper moving and sales coming in. After shifting the data by six months, we had a solid indication of what the market would do over the next six months. (NOTE: The graphs above are an example based on a real experience. The numbers shown are not accurate and are meant for representation only and have been changed to maintain confidentiality.)

Go back to your factory areas and cost drivers from steps 1 and 2 to help you think of potential correlations that impact your results. If you find a Pareto rule, that could be a good area to focus on first to see what drives change. Regression and visualization are two great techniques to help you identify and analyze the data which set us up for the next part of Step 3, making projections and predictions.

Before we move on, I want to touch on data analysis from the physical sensors and devices you may be using. The most common and simple approach to using sensors with microcontrollers like we mentioned (Arduino microcontrollers, buttons, motors, scanners, etc.) is to use what's called an Analog to Digital Converter (ADC). An ADC is built into the Arduino microcontrollers, and it converts a voltage range that it is reading into an integer range. So, if the microcontroller is set up to read voltages between 0 and 5 Volts it will convert this into a range between 0 and 1023.

If we go back to our example of using a button to log start and stop times, when the button is pressed it can shift the voltage being read by the microcontroller which will change the numerical value being logged by the microcontroller. Then, code can be written on the microcontroller that looks for this change in value which will then trigger it to log the current date, time, and employee info. For example, if the ADC is reading 3 volts (614) normally and then drops to 1 volt once the button is pressed (205), it will log the data every time the voltage drops from 3 volts to 1 volt.

Taking this further, you could have the start button shift the voltage to one value, say 1 volt, and the stop button to another value such as 2 volts using what's called a voltage divider circuit. Then, the code will be able to distinguish between the specific button being used and will log the information correctly associated with the right category of start or stop by looking for the exact ADC numerical value associated with 1 and 2 volts.

In general, when collecting data from physical devices, typically values are timestamped and exported to a database or CSV. You can output raw ADC values (0-1023) that will be converted later into meaningful metrics (e.g., on/off events, temperatures, start/stop times) or have the microcontroller process the ADC values first via code, exporting a cleaned and converted list of data ready for analysis such as timestamped 'start'/'stop' events, specific temperatures, and more.

This balance of compute, control, and cost of microcontrollers is critical in edge computing strategy. Too much computing at the edge and you risk losing control because the device spends too much time crunching code causing it to miss important control commands. However, not enough computing results in very raw data output that needs to be excessively cleaned and organized later which can add considerable time and confusion to your analysis. You can deploy powerful edge devices that compute and control well, but these typically cost more than basic microcontrollers doing one or two tasks.

There are a lot of options when it comes to collecting and analyzing physical events because devices can be connected together. So, you may have a central database that collects all information from several basic inexpensive devices each recording several physical events. Spend time beforehand thinking about how you want to connect devices, assign tasks, and collect data and always remember to start small and pilot one process first before scaling out.

This is a very simple example to explain how physical actions can be turned into data and then analyzed, but you'd be surprised by how many applications can use this concept of voltage change. For more help and walkthroughs on this topic and more, check out my website or YouTube channel. Now that we've covered analysis of our known data, we can pivot to look forward by making powerful predictions and projections that turn data into action.

Make Predictions / Projections

Predictions and projections are the trickiest part of data analysis because no one knows what the future has in store. To combat this uncertainty, I want to cover one simple tool that you can use quickly that pairs extremely well with all your new data analysis. Later, we'll cover some more advanced strategies to improve predictions such as logistic regression when we execute externally in step 6. For now, though, let's talk about the data dashboard.

One overlooked cause of failure to achieve projections is keeping everyone aligned with the plan over time. I know from experience, everyone gets distracted as the year goes on. Priorities shift and people sometimes forget what the most important tasks are. This is where a dashboard can really help consistently achieve projections. A dashboard is a great tool to summarize the data, the strategy, specific action items, and key goals. It shows where you've been and where you are going all in one easy to read display.

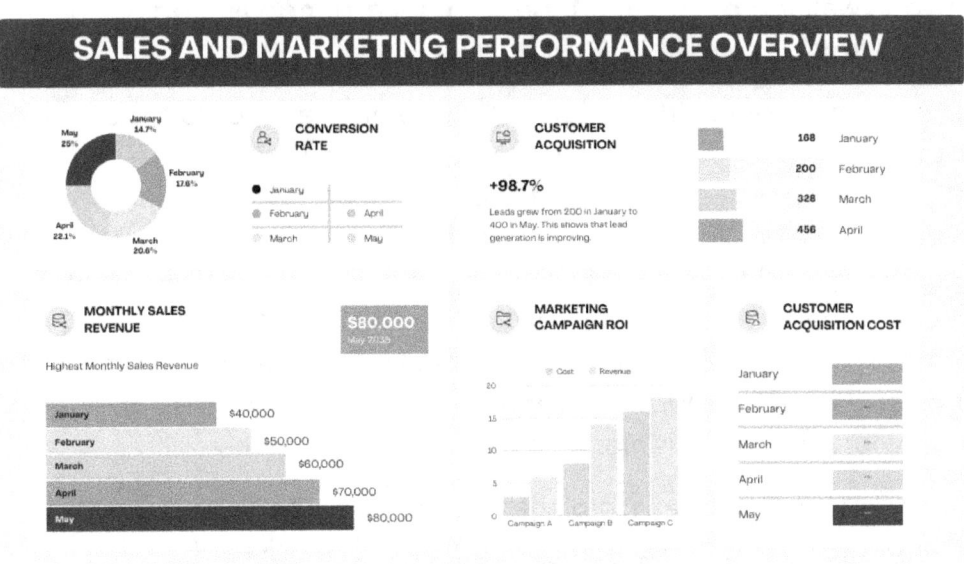

Figure 13: Data Dashboard

One thing I like to do is draw trend lines on graphs to show the progress needed over time. This drives home the reality of what needs to be done day-by-day or week-by-week. For example, if you're trying to cut 12% of costs of goods sold over the next 6 months, that's 2% per month. Can you identify areas to get the 2% cost reduction each month? What actions need to be taken over the next month to make it happen? A good dashboard will drive home the urgency and can also be a source of pride for the team (everyone likes to see the impressive results achieved).

So, when considering a dashboard to be built, start simple. I have often included a summary sheet up front in my analytics Excel files that contain all the data visualizations and summaries from the data analysis. These data visualizations are directly tied to the various inputs and macros, so they update automatically when new data is added. Then, I take this first sheet and turn it into a PDF, print it on 11x14 paper landscape orientation, and then tape it outside of my office once a month. It's a remarkably straightforward way for anyone walking by to see our performance, our goals, and our plans on how to get there.

If you want to get a little fancier, RStudio has dashboard tools built into it such that you can create an html-based R Markdown dashboard that users can log into on their computers to see virtually. This has added benefits in that you can easily update it more frequently (once a day, once a month, or in real time). Also, if your company uses a CRM tool like Salesforce or Microsoft Dynamics, they have their own dashboard capabilities built in (PowerBI for example), however remember to verify the CRM output with your own analysis along the way.

Once you have a system in place to analyze, summarize, and display your data, the fun part can begin. Develop a mindset of experimentation and use the dashboard to track progress. Write a hypothesis on the dashboard with your strategy. Track it over time. If something isn't right, the data will tell you and you can adjust your hypothesis and move on. Fail, adapt, improve. Eventually, this process will become second nature, you'll have a firm grasp on your business performance, and you'll have a better understanding and intuition of what's going to work and what won't. Remember to keep an open mind and let the data guide your way.

With data verified, performance dissected, and projections on the dashboard, you're armed to attack inefficiencies head-on. Next, Step 4: Map and Optimize Processes. Here we will build an agile system that ensures your team members are working on the right tasks at the right time. These improved processes will build on steps 1-3 and finalize the foundation we'll use to act on in steps 5-7. In short, it completes the bridge between "what's wrong" to "how we fix it" unleashing lean operations that scale to victory.

Quick Win: Build a simple Excel dashboard for one cost driver or one business goal, aiming for 5-10% monthly improvements. Use the Supplier Management template with macros for category summaries and audit tracking as a template and integrate it into your dashboard for a quick win.

Risks and Mitigations: Over-relying on incomplete data or getting bogged down in complexity. Mitigate by piloting on one dataset, cross-checking with manual audits first, and always tying back to your factory categories. This reduces waste and keeps things sustainable.

Step 4:

Map and Optimize Processes

Step 4: Map and Optimize Processes

We're beginning the transition from how the business performs (Plan) to "how do we make it better?" (Act). In Steps 1-3 we created the data foundation, and in Step 4 we're going to expand it by incorporating new and improved business processes that are aligned and ready to scale.

Put simply, businesses need processes to scale small wins into big success. Good business processes enable people to grow and adapt naturally and independently but with guidance.

Where does guidance come from?

It comes from culture, it comes from people, and it comes from process. People will tend to default to process first because it's what's "on paper". Process is what shows up on a job description as essential duties. It's what gets discussed during a performance review.

Culture and hiring are a big part of what drives process and results too, however developing a positive culture and hiring good people are complicated and nuanced skills that don't fall into the scope of this book. So, we'll focus on our main controllable, setting up clear and effective processes.

One distinction I want to make with my unique approach: I'm not implying that a process is needed for everything. Quite the opposite. I've seen firsthand how this type of micromanagement-by-form leads to slow movement and bureaucracy stifling progress and growth.

Processes must be managed in a way that allows for freedom and flexibility to avoid stifling explosive growth. So, let's break down how to achieve this.

Set Guidance using Decentralized Leadership

I've tried a lot of management methods over my 15+ year career, but the best one I've used to effectively manage high-performing teams is a decentralized leadership approach that empowers everyone to lead at their level.

If the goal is to establish processes that adapt, provide guidance, and are flexible we need to match that goal with a management style that supports it. A decentralized system promotes flexibility, accountability, and guidance which is exactly why it makes a good pair with the type of scalable processes we will establish in this step.

Additionally, this management style avoids micromanagement which gives executives time for high-level strategy without bogging down in task details. For more on why this is true, read my article at:

https://www.jasonunlocked.com/blogs/articles/mastering-the-executive-mindset-for-tech-innovation-and-business-growth.

Here are the key points to this approach based on my experience of building high performing teams that regularly beat targets:

- Define the overall purpose, desired outcome, and goals of the initiative.
- Keep the messaging simple so everyone from top executives to front-line staff can understand it.
- Clarify how each person's role fits into the bigger picture.

Define the overall purpose, desired outcome, and goals:

When collaborating with your teams, kick off the strategy or initiative with only the high-level purpose, goals, and outcomes. Let your team lead the discussion on how to get there. If they need some encouragement or help, you can step in, but the key is to have the team take over responsibility as quickly as possible and avoid micromanagement.

Keep the messaging simple:

Transform the strategy's goals, steps, and responsibilities into crystal clear language that anyone can understand. Everyone needs to be able to comprehend it up and down the chain. This removes doubt, uncertainty, and delays and keeps everyone focused on the main objectives.

Clarify each person's role:

Make sure that every task has a person assigned to it. This enables teams to pivot and work faster because they can focus on individual assignments without getting slowed down by larger strategy problems or worrying about who's doing what.

For flexibility and speed, empower team members to take initiative, adapt, and strategize on their own within the guidelines of the top-level objectives.

For guidance, check in regularly to make sure the team is heading in the right direction but try to stay out of the process as much as possible.

In fact, I've occasionally let team members make minor mistakes that I could see coming so they could learn firsthand how to strategize and execute better. After all, experience is often the best teacher, and everyone makes mistakes. Don't try to set up a perfect process, set up a process that enables growth, accountability, and flexibility. Just be sure to jump in for major problems and use this tactic strategically. Don't let it set back the team in a major way.

Keep this approach in mind as we move forward because next, we're going to pair it with a foundational understanding of how your company really works. Together, these key concepts form the blueprint for creating world-class processes that not only look good but deliver outstanding results as well.

Clean Out the Garbage, Prepare for Growth

Now that we've adapted our leadership style, let's upgrade our processes to match. In my experience, there are always two sets of processes within any company. First, there are the official processes that are fully documented and used for auditing purposes and certifications. Second, there are the unofficial processes that people use to achieve results. Sometimes they are the same, often they are not.

To create effective processes that scale, we need to understand the actual processes people use to get work done, not just what's on paper. Without this understanding, building on Step 3's insights will become impossible. You'll be creating and assigning tasks that don't align with how the work gets done. In my experience, this mismatch between processes is a huge contributor to the 95% of tech initiatives that fail. It's also a huge contributor to the massive spending on custom software solutions and consultants.

It might be confusing to think that people work outside the official processes, or that executives don't fully understand how work gets done within their teams. But, I've seen firsthand the disconnects that can occur. Usually, it's not intentional. One reason these hidden processes aren't recognized is because people are overwhelmed with tasks and can't properly document and update their processes.

Another is that process mapping typically falls in a grey area. People doing the work aren't assigned the task to update processes. Usually there is another team managing the quality and process system, but they're not the ones doing the work and so they don't catch all the nuances of the processes.

Also, discovering hidden processes can be difficult because people don't typically admit that they work outside the official process. They fear it might lead to bad outcomes such as failed audits or bad performance reviews.

The reason I'm spending so much time on this is to drive home the point that as executives, we need to get to the heart of the matter. And, because we set the process and have a large say in company culture, it's important for us to create an environment where people feel comfortable talking about the good and the bad aspects of their work. Without this, the company will always struggle to adapt and improve no matter how great the insights and plans are.

Now, going outside the official process is not always a bad thing, but it needs to be identified. If the hidden process is leading to superior results, it should be implemented and shared with the rest of the team to multiply the success.

So, let's talk about how to discover the real processes. The first thing we can do as executives is to build trust which will open communication. I've found you can take two approaches:

1) Explain how you're looking for honest feedback on how to do things. People may or may not open up about this, but if you are consistent and incentivize feedback it can eventually work. Position yourself as the student learning from the team.

2) Explain how you're exploring a new decentralized structure like discussed previously. You want to empower the team to greater success and to do that, you need to know the full picture. This tends to get better and faster results in my experience. People will open up when they know that there's a real chance to remove the red tape and make the situation better.

What works for you may be a combination of the two, or a different tactic. A good leader will know the best way to communicate with their team, but remember the main goal is to build trust so the team opens up to you about how things really work.

One other good technique is to spend some time in person in each division and within each function shadowing team members. Be open and inquisitive, ask questions, and have them teach you the process. Again, it will take nuance (people don't like it when someone is looking over their shoulder) but if done correctly and persistently it will work.

Now, there are two major outcomes from discovering the real work being done. First, if you have established processes, take notes (or have your team take notes) and then compare them to the established processes. Find the good and bad deviations and change as necessary.

Second, if you don't have established processes (or if you find new processes) use this as an opportunity to start documenting. This is important because it limits your ability to scale if not done correctly (steps 5-7).

I can't emphasize this enough, process mapping is not about restricting activity, checking boxes, or creating more work for people to do. It's about the ability to scale. I don't care how talented your best people are, there are only so many of them and at some point, you must have a system where the second and third-tier team members can contribute. This is what effective process mapping does, it teaches and guides your people to get the best results in the shortest amount of time for all tiers of performers.

When starting this step, be sure to reference your factory model and the data you collected in steps 1-3 to set priority of which process to review first. You've identified, gathered, and analyzed company data to find the most important core processes, inputs, and outputs so start there first. Doing this will help you get the 10-20% gains quickly. Then, you can move down the chain to the mid and lower impact processes over time for further refinement and improvements.

Now, sometimes absolute control is needed (for auditing purposes or safety), however the goal is to develop a system of process mapping that is fast, flexible, and empowers your team to maximum success. And, this is where we can leverage technology to help.

Leverage Technology for Explosive Growth

As you begin to map out your processes, you can leverage technology to reduce the workload for your teams. There are two great low-cost software apps that can help in this regard:

1) **Lucidchart:** A very intuitive tool for building virtual dashboards for flow charts and processes. It can accept time / metric data into the process for real-time tracking and process refinement. Teams can work together on a process. Lucidchart shines because its real-time collaboration eliminates the need for endless meetings, letting you overlay Step 3 data like time metrics to reveal bottlenecks instantly.

2) **UiPath (also Zapier):** An automation tool that allows team members to easily set up connections and tasks for redundant, low value add processes. With UiPath, you can automate repetitive steps post-mapping, integrating with tools like Excel for seamless workflows.

A big reason to use technology at this point is in how it compares the actual vs. official processes you have. Technology like Lucidchart makes it easier to compare the results of two sets of processes to see which one is better and how it affects other areas of the company. It also makes updates easier so that you can iterate quickly to get to the best possible process. And, because team members can work together, it supports the decentralized structure talked about earlier.

Here's an example of what this combination of data and process analysis can do. It started with a question I had one day: Where do people really spend their time during the day? At one point in my career, I was responsible for sales, but I discovered that I was spending ~30% of my time on data entry and management. So, I started to look at the process and figured out how to shift my work to prioritize the critical tasks and leave the less critical (such as data entry) later when I had down time.

This gets even more powerful when done as a team. Imagine if you had a centralized place where people could input actual times (Like Lucidchart). It eliminates the need to even discuss it. You can see where the bottlenecks are and work to refine them. This keeps your team focused on the critical tasks (like generating revenue).

Here's how technology can streamline a sales process based on a real process from my experience. This is a process to convert a customer conversation into a sale:

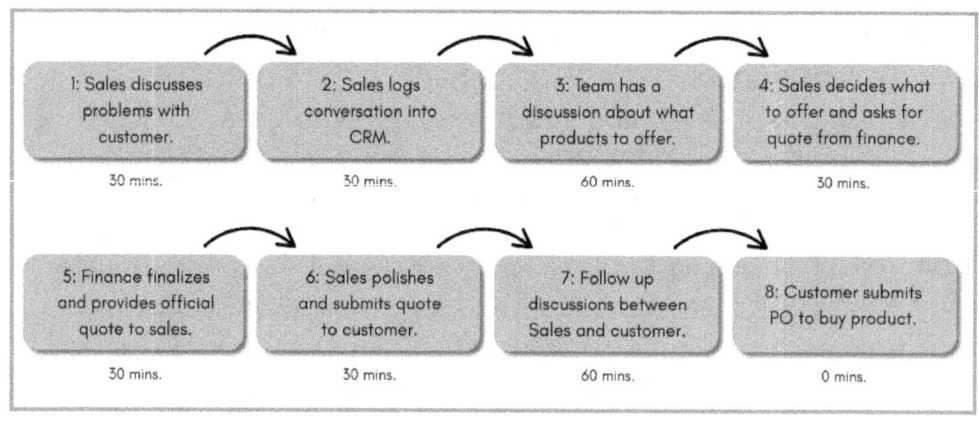

Figure 14: Process Mapping Before

Now, here's the revised process using technology:

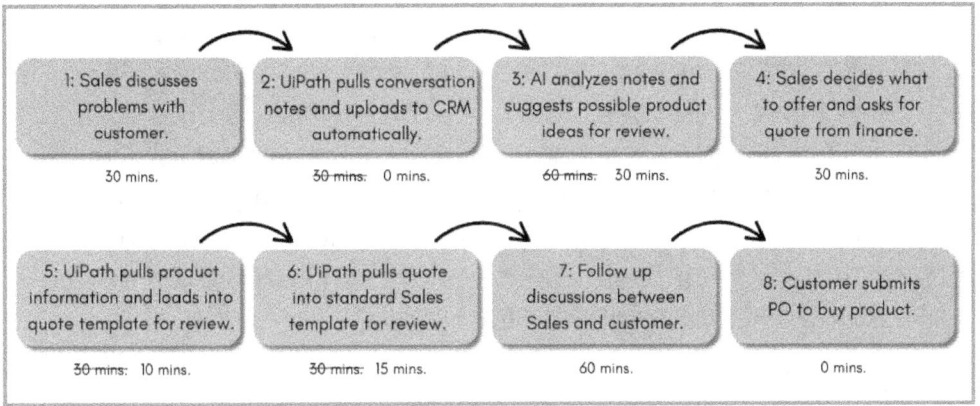

Figure 15: Process Mapping After

Original process: 270 minutes (4.5 hours)

New Process: 175 minutes (3 hours)

Process Savings: 35% Savings (1.5 hours!)

Figure 16: Process Mapping Summary

This is just a simple example. There are many more processes that you can simplify and automate in this way. For more examples of how to use AI in operations, check out McKinsey's 2025 insights highlighting the use of gen AI and how it can add to process maps to boost efficiency and decision-making in supply chains:

https://www.mckinsey.com/capabilities/operations/our-insights/generative-ai-in-operations

Encourage Ownership for Fast Results

One final tip from my experience when refining your process mapping is to leverage the effects that come from ownership of tasks. A big part of discovering the real process is to start a dialogue around change and improvement. Then, to make process mapping even more powerful, involve the team and encourage them to take ownership of the process as you finalize the new way of doing things. Remember our focus on decentralized leadership? It plays perfectly into this strategy and here's why:

The people doing the work need to take ownership of the process. They need to have a heavy say in creating it. Without this, people will work against or outside the process and point fingers. This is the exact reason I use the decentralized leadership strategy discussed earlier. When the team can strategize at lower levels, it automatically encourages ownership of the process which leads people to accept the changes faster and work together toward finding the best solutions.

This works because it's human nature to avoid being inconsistent and unreliable. Culture, especially business culture, heavily encourages consistency and reliability which turns into credibility and trust. So, when your team attaches their own names to tasks, they will go the extra mile to complete them successfully. Use this when defining your processes.

So, in summary, to get the best results with process mapping use decentralized leadership, find the real processes, incorporate technology, and encourage ownership. If you follow these principles, you will establish processes that are fast, flexible, and accurate. Processes that empower your people to use the insights found in step 3 to maximum effect and quickly. These processes will enable your company to scale fast and grow while still fitting within the overall guidance and strategy of the company.

When done correctly, you can do more with less while eliminating waste and improving culture along the way. At one point, I was given responsibility for all North America for several market segments, and I was able to manage it with a small team because of techniques and tools like this.

And, most importantly, everyone loved it because it gave them the control and freedom to do what they knew would produce results instead of the time-wasting tasks that everyone hated. My team became a magnet for top performers because they saw the freedom and results that came from this framework. It enabled the front-line workers to generate results for our customers, and it created a system that ensured our processes were being updated properly so others could benefit from the work.

Now, before we move on, if you want more information on process mapping in general, this is a great resource to learn more of the "nuts and bolts" of process mapping itself:

https://www.lean.org/lexicon-terms/value-stream-mapping/

Let's take a step back and see how far we've come. In Step 1 we identified and prioritized cost drivers and organized our business into factory segments for easy tracking. In Step 2, we deployed data gathering tools. In Step 3, we analyzed the data and set up dashboards to help us make sense of the noise. In Step 4, we've prioritized and mapped our processes to complete our new foundation of data and understanding. Now we can transition to action in the next steps of the process.

In the next step, Step 5, we're going to focus on how to implement changes within the company factory. Step 6 will focus on external actions, and Step 7 will tie it all together for a scalable strategy that delivers quick ROI and sets your team up for sustainable growth.

Quick Win: Shadow one team for a day to map a core workflow using Lucidchart. Aim for 10% immediate efficiency gains by spotting and automating one bottleneck, like data entry.

Risks and Mitigations: Over-mapping can lead to bureaucracy. Mitigate by piloting on one high-impact process from Step 3 data. Team resistance to sharing real deviations? Build trust persistently through the approaches above. Tech over-reliance early on wastes resources (remember, 95% of AI pilots fail without basics). Start manual, then layer in tools.

Step 5:

Implement Internally

Step 5: Implement Internally

In this step we're going to start making changes to your business based on the data you gathered and analyzed in steps 2 and 3, the cost drivers you identified in Step 1, and the processes you mapped out in Step 4.

If you've followed along with the process, it should start to become clear at this point which areas of the business you need to focus on first. The changes you decide to make will be unique to your situation, and if you implement them with your new data collection and analysis systems in place, you should see improvements fast (or know where you need to pivot quickly).

Now, this step is not going to be an extensive list of generic advice on how to implement change in business. It's impossible to cover all the unique variables and outcomes in one book.

So instead, I'm going to focus on two key areas of internal change that will help you sustain long-term success with your new initiatives and foundation. If you find that you need specific help with your problems, you can always reach out for a consultation session at:

https://www.jasonunlocked.com/pages/consulting

Or, head over to my YouTube channel at:

https://www.youtube.com/@JasonBUnlocked

The two major areas of the factory business model we'll dive into are the data infrastructure as well as executive oversight with an emphasis on change management. The changes in these two areas will ensure that the company is fully supported with the right hardware, software, and oversight to implement your changes successfully and sustainably.

Data Infrastructure and Management

First let's start with a critical area of the infrastructure that supports all other internal areas of a business, the data and IT infrastructure. This part of the business feeds directly into the core processes and drivers of your business factory and enables all the advanced data collection and analysis that you'll be doing as you move forward. So, let's dive in.

Data is the currency of business. This book introduces a lot of techniques that rely on sound data management principles to implement and scale over time. So, it's critical that you spend time thinking about how the company manages data. The end goal is to ensure the business is ready to accept and store data securely and efficiently while still granting access to as many people as possible within the business so they can find new trends and efficiencies.

Most companies have broad departments covering this area. For example, the IT department is often responsible for anything related to technology and data management, along with everything else like computers, phones, and printers.

Having one, or a few, departments responsible for everything is no longer an effective solution to achieve lasting success. Business relies on data; you must take a more detailed approach.

I've found that this area needs to be divided into five sections which are:

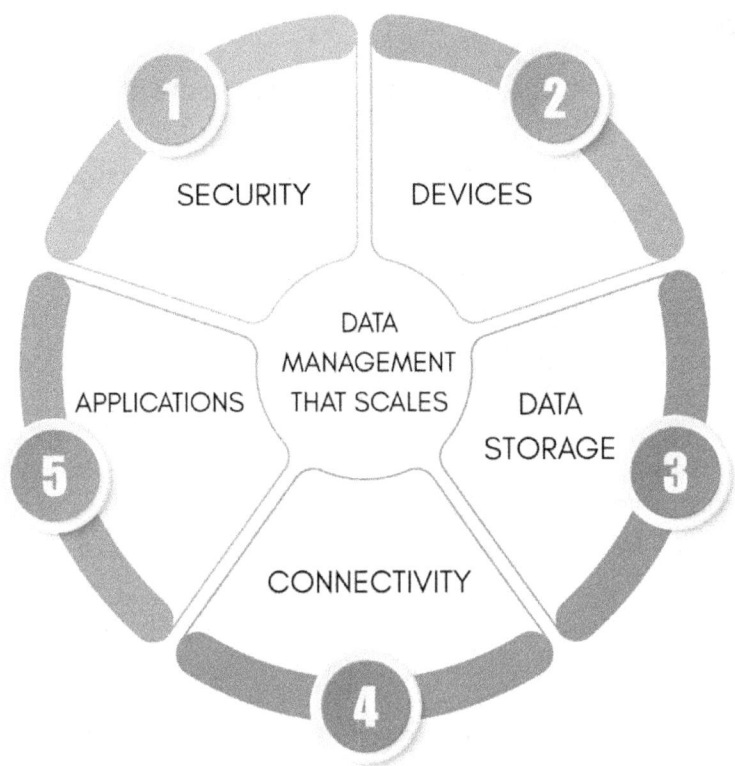

Figure 17: IT Management that Scales

Depending on your size and resources, you may have individual teams working in each area, or you may have people working in multiple areas. However, the work should still be split into clear areas of responsibility. For example, one person manages Applications and Connectivity, another handles Data Storage and Devices, etc.

The purpose of this IT re-org is to prepare your business to be fast and flexible, to scale while remaining secure. To achieve lasting success with your new data collection and process improvements, you need world-class IT infrastructure. Let's discuss each area in greater detail so you can find the right fit for your company.

Security

This is an overarching area monitoring all aspects of the company's digital infrastructure. The main responsibility of this team is to ensure security while enabling proper access to the right people as much as possible.

Attacks are becoming more advanced and more frequent, but that doesn't mean you should lock everything down. An effective way to support security and access is to ensure that there are processes set up for people to apply to access data. Your security team can set up training on how to do it safely and securely.

Now, the actual security solutions you choose will be up to you and your team but remember both sides of the mandate: security with proper access. If you lock down too much, people won't have access to the data they need to make timely decisions. Finding the right balance can be tricky, here are some warning signs to look out for:

In my experience, most companies default to full lockdown with very few people having access to data. I've seen processes take days to relay critical information from IT to the right people. And, if the data set is wrong, the process starts all over. The problem comes from a mismatch in skill sets. IT people are typically not data experts, and the data experts don't have IT knowledge and access. It creates an endless loop of wasted time and resources.

For example, at one point in my career the only tool available for data analysis was Excel due to restrictions on third party applications like Python or RStudio. Once I started analyzing tens of thousands of lines of data, Excel started to slow down and lock up. Processes that should have taken minutes started to eat up hours of time. I had to re-code the whole program several times to find ways around these problems which would have been unnecessary with a different tool. If there had been a security process to use new programs while protecting data, it could have saved hours of time while enabling better results.

Most executives are shocked when team members tell them that simple tasks like data collection, or communication of data, can take days or weeks. It does seem silly at face value; one person just needs to communicate with someone else. But believe me, it can happen if the processes and infrastructure are not set up correctly.

The balance of security with accessibility should be a top priority for success and scalability. Without it, taking advantage of the work done in steps 1-4 will be extremely hard.

Here's something else to consider with security: If systems go down and you're being hacked or disrupted, you want to be able to pick up the phone and call your own team of experts right away.

Millions of dollars are lost every year because companies do not have the proper security expertise in place. When a security event does happen, organizations that aren't prepared must scramble to bring in third party experts to figure out what happened. But, by this point it's too late and they're going to pay for it in legal fees, lost business, and lost reputation.

But this is just the tip of the iceberg. The takeaway is to give this area its proper attention and always consult with professionals about your unique situation.

For more information on best practices on balancing security with access, like adopting least privilege and regular reviews, visit:

https://pathlock.com/learn/user-access-controls-11-best-practices-for-businesses/

Devices

The connected devices that the company will use going forward will be much more than the basics of the past (computer, phone, maybe a tablet). Now, we're talking about IoT with sensors, microcontrollers, AI enabled edge computing. It can quickly overwhelm a company (especially when scaling) when the number of devices being managed starts to grow.

This is why I recommend setting up a dedicated team managing this area of the business. It's not just about adding in new devices. Over time, the team will have to think about maintenance and replacement along with new trends and requirements that come up. And that's not all that this team can do.

This team can also branch out into more advanced areas like PCB (Printed Circuit Board) design, PLC (Programmable Logic Controller) design, and even component design. If done correctly, you could find this team creating new business verticals of proprietary devices that work for the company and that can be sold outside the company. If your team is developing industry leading device solutions, there's a good chance others will want to buy them.

Data storage

A data storage plan is critical to ensure security, accessibility, and to keep costs as low as possible. Some key questions to consider:

- Are you storing data locally or in the cloud?
- Will data be accessed frequently, or infrequently?
- Do you have several branches around world that need access, or just a few local locations?

The answers to these questions can drastically affect your storage choices and the amount of money you pay to store data. It's a continual process to monitor the flow of data to ensure it's always available and stored in the best manner for your business.

For example, some data will not be used frequently such as compliance information, past transactions, charter documents, and more. This is information that you need to keep but will certainly not look at unless there is an audit or contract dispute. For this data set, you can choose a cold storage option and pay very little to store it.

On the other hand, the data that makes up the backbone of your analytics in steps 2 and 3 will need to be accessed quickly and frequently. Therefore, you will need a more responsive storage option and will pay more for it. But, by producing an overall strategy that balances performance and cost, you can offset some storage costs while ensuring quick and reliable access to the data you need.

For more information on data strategies, see my article at:

https://www.jasonunlocked.com/blogs/articles/unlocking-smarter-execution-how-executives-can-rethink-data-access-for-better-decisions

Connectivity

With more devices comes more digital communication. How devices connect within your business will require a separate plan and a team with unique skills.

Here, topics like bandwidth, spectrum, and language are critical. You will want to ensure broad and stable coverage without interfering with other systems. Again, we're moving past just the basics of phones, laptops, and tablets. Each device you add in will require careful thought to ensure it doesn't take down other systems.

This becomes critical when you start to integrate factory sensors into other systems like CRM and ERP systems. In the same way that we used CSV files to translate between different programs for analysis, hardware sensors also need a common communication form to ensure data and commands are passed successfully. There are all kinds of communication protocols depending on devices and strategies such as Bluetooth, Wi-Fi with Web Sockets and TCP/IP protocol, Serial UART, CAN/LIN, HTTP/MQTT, I2C, NFC and more.

And, if a connectivity problem arises, you need a dedicated team ready to address it head on because it could take down your whole factory. This is why it gets its own dedicated team.

Applications

This team is not just responsible for managing existing enterprise software solutions (like Microsoft Office), it should also be used to generate new software for the company.

The people working in this group should have coding knowledge and be interested in finding the best software solutions for your problems, whether that's an existing solution like Salesforce or Oracle, or building a new custom solution. This enables your teams to move quickly by having dedicated resources to help them achieve their data goals and strategies.

Even most enterprise software programs need code adjustments to work perfectly for each company. This is another area where I've seen companies pay millions of dollars to specialists to come in and ensure their latest version of ERP software works correctly.

The advantage of building this team is to mitigate these costs and you might even find a new business opportunity by spinning off some of your software solutions into standalone companies. This team will be critical for your growth and scaling success, so break it out into a separate team.

A few key points to help wrap up this section:

Be careful about the silo effect with this new structure. The management team overseeing this new organization must ensure that teams are working together effectively and that the output matches the high-level strategy of the organization. Remember the decentralized leadership methodology. Empower the teams to execute at their level while ensuring proper high-level oversight.

As with other areas, start simple and start small with these changes. An IT re-org is not too difficult, but changing communication, storage, and connectivity can have severe consequences if done incorrectly.

So, the first step should be to organize what you already have in place. Set up the new team areas and then take inventory of how your business runs now to set a baseline foundation. If you find critical holes or vulnerabilities in this process, address those immediately.

When moving forward, establish a playground / quarantine process where new devices, data, and software can be tested and perfected outside the main processes of the business before integrating them into the production processes of the business.

For example, you might add a new sensor system to one product line or add one new software program to one team within the business first. Ensure it runs smoothly without introducing new risk into the system. Then, broaden out to other areas. In this way, you can set up a technology flywheel within your company that lives and breathes continuous improvement, exploration, and explosive growth opportunities while minimizing risk as you scale.

Before we move on to step 6, I want to talk about another key area when implementing change internally: The executive management team. It's critical that the executive team is deeply involved in each area of change, not just the data and IT, but all areas.

Executive Oversight and Change Management

Change is hard for most people. Some will resist while others will be eager to take on the new challenges. The best way to overcome the pushbacks while encouraging those eager is to ensure the executive team takes an active role in the process. Now, I'm not advising them to jump in and start doing the work, but they must show up so people know it's important. Appearances and attention matter here.

Some strategies for this:

- **Set up regular high-level meetings** where the executive team hears status updates, new initiatives, wins and losses, etc. When senior leaders recognize the team, they will take it seriously.
- **Showcase wins in company meetings.** If you have town hall meetings or monthly team meetings, showcase some of the wins and have the teams explain what happened. This will create a positive culture around the change and get your teams working together.
- **Assign areas to executives and rotate them.** For example, for one year an executive might oversee the IT changes. Then, for the next they could move on to factory operations. This ensures that the executive team is getting broad exposure to the business, and it ensures that it keeps the process fresh for everyone. The team gets new leadership that may take things in a new direction, the executive gets knowledge of the larger business that helps them make better high-level decisions. Treat each assignment like a sprint. For example: "For one year, we've got to achieve X, Y, and Z". This will create excitement and keep projects moving forward effectively.

Another key aspect of executive involvement: cross-functional management. A lot of these changes will be made across departments, therefore, having the executive team engaged will ensure smooth cooperation and remove some of the political barriers and red tape that can come up when people start going across silos.

Finally, don't forget to celebrate the wins. Constant change and action can burn out anyone. Take time to celebrate wins in the form of concrete rewards (bonuses, time off, etc.). This will draw a clear line in the sand that people can move forward from. It creates a culture where people look for wins eagerly, close them out, and then move on to the next one.

Let's take a step back and see how far we've come. In Steps 1-4, we identified, gathered, analyzed, and mapped processes. Now we've laid the foundation to implement internal changes like an IT re-org to deliver lasting success in your factory's new updated core processes.

In Step 6, we'll extend these wins externally to suppliers, sales, and markets for bigger impact before synthesizing it all in Step 7 for a scalable lean strategy.

> **Quick Wins:** Run an IT baseline audit in one area, like Data Storage, using a simple spreadsheet to spot quick cost savings (e.g., switching rarely accessed data to cheaper cloud options for 5-10% reductions in weeks).

> **Risks and Mitigations:** Over-specialization in teams could create silos and reduce cross-communication leading to inefficiencies like Security clashing with Applications. Mitigate this with regular executive-led meetings to keep everyone aligned. For smaller businesses, dedicated teams might strain resources with hiring costs. Start with multi-role assignments as mentioned, and scale as you grow.

Step 6:

Extend Externally

Step 6: Extend Externally

As internal operations begin to improve and you build out the new foundation of the business, start to transition these principles to the external ecosystem to amplify gains further.

Technology plays a vital role in key external relationships such as the customers, suppliers, and logistics partners. Here's how you can leverage it to slash costs and streamline your external processes.

First, take inventory of what you already use. There are some great tools out there that help with external partners, and you probably already use some. whether it's a CRM system, an ERP / MRP system, social platforms, or even AI enabled programs. Even though these programs can be very powerful as is, it's always useful to build out your own simple tools when possible as a check.

This step will discuss some practical steps to fuse your new systems and skills with your existing systems to create synergies and reduce costs. Best case, you might end up using your own solutions or customizing the existing solutions to better fit your processes which can save millions in the long run. Let's start Step 6 by looking at how we can improve customer relationships.

Customer Relationships

Most aspects of customer relationship management (CRM) systems involve conversations and data entry and management. Generative AI systems work amazingly well to help offload the mundane tasks in this area. Some great ways to use it include:

- Collecting and summarizing meetings notes.
- Generating action items and implementation plans.
- Generating ideas, evaluating, and ranking plans.
- Creating schedules and next steps based on notes.

Technology is rapidly advancing and now there are some tools that are working AI directly into the CRM process. But, if these programs are too expensive and complicated to implement, or if you're having AI issues with safety, security, and your data, you can work around these problems using methods like the one that follows.

After any important interaction you have, grab your phone, and open a note file or start an email to yourself. Then, using the speech to text option (available in most phones these days), dictate notes on everything you remember from the conversation. Form or polish do not matter; simply regurgitate everything you remember along with any other ideas or actions that pop into your head while you speak. Let it be a stream of thought that includes what happened and what you want to do with it going forward.

Then, email the document to yourself and copy paste it into an AI system after removing or scrubbing any protected or confidential information. Ask the AI to organize, summarize, and add context to the notes. If you use a CRM system that allows you to upload data, ask the AI to structure the notes into the right sections with headers so the notes can be uploaded directly into the CRM system. In this way, you can create a powerful system that offloads most of the work when it comes to customers and data management.

A word of caution: Be careful how you combine proprietary data with third party services like AI. If you are using an enterprise tool like Microsoft Office, it may have Copilot built in that is already set up to protect your data. If you are copy / pasting into an AI system, be sure to anonymize and strip out proprietary data first. This is not an exhaustive list, always consult experts for your unique situation. Ideally this is something that your Security team takes on as mentioned before. They can establish the proper tools and processes to safely transfer the information whether you are using a phone with notes or a more formal enterprise solution like Microsoft Office.

Markets and Go-to-Market Strategy

Now let's pivot to markets and explore how technology can enhance your go-to-market strategy. Simple tools like Excel are great starting points, as we've seen in Steps 2 and 3, but for more advanced customer targeting, a free tool like RStudio takes things to the next level. It's designed specifically for this kind of analysis and handles large datasets effortlessly. Here's how.

In Steps 2 and 3 we explored linear regression to find correlations. There is another type of regression, logistic regression, which helps identify new demographics to target based on your most successful customers to improve your go-to-market strategy. While linear regression spots trends in numbers like sales growth, logistic regression predicts yes/no decisions.

In the world of customer analysis, logistic regression is a simple yet powerful tool for turning raw data into actionable insights. Imagine it as a smart filter that predicts "yes" or "no" outcomes, like whether a potential customer is likely to buy your product based on simple traits such as age, income, or past purchases.

Once set up, this type of model will analyze thousands, or even tens of thousands, of lines of data on new potential customers, compare them to your existing best customer list, and then assign a probability score to each new prospect showing how strongly they match. In this way, you can sift through large amounts of potential to find the best prospects and then assign your team efficiently to target those new prospects.

Start the logistic regression process by organizing your existing customer base in a spreadsheet (Excel or CSV works fine). List each customer's details in columns (e.g., "Age," "Annual Spend," "Bought Product?"). Then, download the free RStudio software and import your data. It's as easy as clicking a button or copying a few lines of simple code, any AI program can walk you through it.

RStudio crunches the numbers in seconds, even with massive datasets. It uses your existing best customer performance data to train a model that will then assign probabilities (say, a 75% chance of purchase) to each new prospect. After training, apply this model to a new list of prospects by loading their data and running a single command. It instantly ranks the best matches without searching through endless spreadsheets. It's quick, scalable, and empowers you to focus on strategy rather than guesswork.

I highly recommend you spend some time exploring analysis like logistic regression. It might seem complicated at first, but RStudio makes it surprisingly accessible even for non-programmers with point-and-click options and ready-made templates. Once you start using it, you'll see how powerful it can be and it's completely free.

When using logistic regression in RStudio, start by focusing on data that defines your ideal customer. These customers are the largest or most profitable ones, which aligns with the 20% driving 80% of results from prior analysis. Dig deeper with demographic details (age, income, or location), which can be fed into logistic regression to predict their fit for your business.

Here's a practical approach to apply this in RStudio:

1. **Identify Your Ideal Customer Profile:** Use insights to pinpoint the key traits of your top 20% customers, such as company size, industry, or demographic segments. Label these as "high-value" (1) and others as "low-value" (0) based on revenue or success metrics.

2. **Gather Broader Data:** Research similar data for prospects across a region like North America. Pull data that matches the traits of your ideal customers (e.g., company sizes or demographic stats from public sources if that's what you used on your ideal customer set).

3. **Train the Logistic Regression Model:** In RStudio, import your data, then use the built-in glm command (a simple one-line function) to fit the model. Specify your predictors (e.g., location, age) and the binary outcome (1/0). The software quickly calculates probabilities, indicating how likely new prospects match your best customers (higher probability means a match is likely). Review the output for insights, like which traits boost the odds. For example, higher income might mean +20% likelihood of purchase.

4. **Target Strategically:** Load your prospect list and apply the model with another quick command. Review the probability scores to prioritize leads. For instance, the model might suggest focusing on the Midwest, South, and Northwest, or older, established companies, allowing you to allocate resources to the highest potential opportunities.

This method leverages logistic regression to emulate the success of your top customers, identifying new prospects with similar profiles in the shortest amount of time. This is how you do more with less and get results faster. Use technology and data to help you find the best targets based on your existing performance.

At first the model might highlight obvious matches, like a customer similar to your current big accounts (Home Depot resembling Lowe's). However, the real power is that RStudio can process massive amounts of data quickly. So, once you have the model finding good predictions on the obvious, move forward and feed it more data. For example, expand beyond North America to include Europe, or refine regions into smaller subsets like individual states. The goal is not to find the obvious, it's to uncover new potential customers you wouldn't have thought of at first.

Additionally, logistic regression in RStudio can help you make projections about changes in your business. For example, if you think new startup companies might be more valuable, re-run the analysis (e.g., reclassifying startups as "high-value" = 1) and see how that shifts your probabilities. Once you have your model set up, experiment with different scenarios to see mathematically how they impact your business. This is not an exact science, but it can speed up your decision-making and deliver a repeatable framework so you're not guessing blindly.

Suppliers, Finance, and Logistics

Let's look at a few other ways we can leverage technology to help with external connections to the business. Working with suppliers and auditing them can be extremely time-consuming for a business, so let's start there.

Suppliers:

I've provided a sample excel file with macros and AI prompts to help you build out a supplier audit system, and now that you have more data and experience using tools like UiPath, you can take this even further by adding more information and automation to the file.

For example, maybe you want to add quality defect information to see which suppliers perform the best. Maybe you want to add a check for signed supplier agreements and / or expiration date of contracts. If you're new to coding, use an AI program to help you implement the code changes. Keep it simple and go step by step to ensure you understand the changes.

For example, don't just tell the AI that you want a macro that can analyze several different metrics of supplier data. Start with one metric (like on time delivery), make the change, ensure it works as expected, and then add another metric. Go one by one to ensure that you don't get overloaded with changes which makes it almost impossible to spot problem areas in the code. Be sure to save copies of your files too in case you need to start over or go back to an earlier version if the code change doesn't work.

Finance:

Another critical area that can benefit from these techniques is the finance team. It can be a real struggle to get customers to pay on time if you operate on a credit system. Sometimes it can be hard to even know which customers have exceeded their allotted time to pay.

And if things get bad and the customer is late, it's important to start tracking things like when they were last contacted, who contacted them, etc. These problems can build up over time if not monitored effectively and you may be missing a lot of money that should be in your bank account.

Use similar macros from the supplier audit file to adjust the data pulled from your finance systems. Instead of bad suppliers, you're filtering for bad customers. You can run a report daily (pull data into Excel and run your macros to clean, categorize, and filter for bad customers) and add tools like UiPath to send out communications to team members on customers that need additional attention.

I've seen this work firsthand. Sometimes all it takes is a phone call from the sales team to get past due invoices paid. The problem is often that the team doesn't know the customer is behind in the first place. So, in this case, a regular report was established that collected key finance information in an Excel file that was shared with the sales team so everyone could review past due customers. And, within a month, we had reduced the past due list by 50%. That kind of change can have a real impact on the business and cash flow.

Logistics:

Another area is shipping / receiving / and general logistics. You can use the same principles as supplier and customer audits for shipping activities to take your operations to the next level. For receiving, you can export data or use tools like UiPath to pull in information from your various shipping partners to a central system.

Maybe your ERP / MRP system already does this, as we've discussed before it never hurts to have a simple check system in place like an Excel file with UiPath and some macros.

Alternatively, you can start to build out your own system if you don't have ERP / MRP capabilities. For shipping, you can automate messages to various stakeholders inside and outside the company. You can send tracking information to customers, create a time stamp in a finance file to start the credit time period, and notify the sales team all automatically using UiPath and Zapier.

Implementing some simple hardware technology like barcode scanners can also help. Start simple with one pick and ship process and have the scanner send updates to a simple database on part status (picked, in shipping, shipped for example). Keep this outside your normal process until you've proved it out, and then gradually move it into your production environment.

For more ideas on how to implement technology with your supply chains, check out McKinsey's 2025 report on gen AI in supply chains showing efficiency gains of 10-20% in workloads through analytics.

https://www.mckinsey.com/capabilities/operations/our-insights/beyond-automation-how-gen-ai-is-reshaping-supply-chains

Remember to start simple, create small simple pilot programs in each area prioritized and categorized based on your findings from steps 1 through 5. Then, gradually expand and add in other information and processes. Before you know it, you'll start to see amazing improvements that prepare you to scale and beat the competition. These external wins set up Step 7's synthesis into a full lean strategy.

And, since you now have experience with dashboards, it's a great idea to set up some dashboards around external metrics as well. Here's an example of what that could look like:

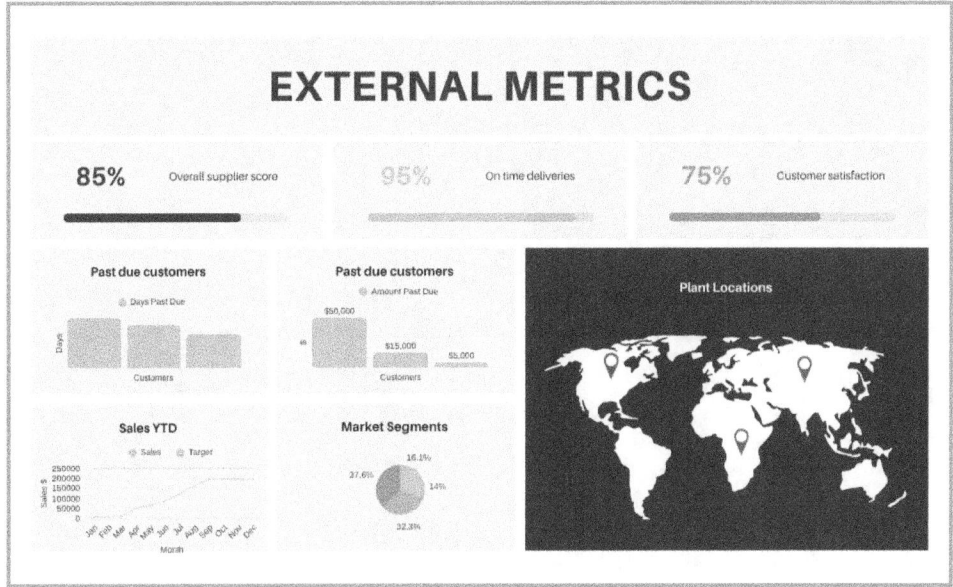

Figure 18: External Data Dashboard

Quick Wins: Pilot an AI-summarized CRM entry to free up sales time right away or run the supplier macro for instant audit flags (e.g., yellow for <3 suppliers per category).

Risks and Mitigations: Over-reliance on AI for customer communications could make things feel impersonal. Mitigate that with quick human reviews before sending.

Step 7:

Unlock Your Operations

Step 7: Unlock Your Operations

Now it's time to zoom out and put all the pieces together into a strategy that you can track and adjust over time. This final step will cover how to use some of the included resources to help you formalize that plan.

A quick tip before we dive in: Think of this strategy as a living, iterative, and lean operations strategy that's scalable, team-owned, and adaptive. This isn't some dusty plan on a shelf nor is it something that is set in stone forever. It's a dynamic engine that empowers your people, tracks real progress, and evolves with your business.

We'll walk through the basics of building your unique strategy with the included Workbook, and recap how it ties everything together.

Again, if you don't have the Workbook head here to download it:

https://www.jasonunlocked.com/pages/bou-resources

Initiative Setup

You've done a lot of excellent work to organize, prioritize, and map out how to implement changes and improvements for your business, now let's spend some time making sure your plans transform into tangible results. The best way to do this is with some high-level structure. That's why we're starting with initiative setup.

An initiative is a project with purpose behind it. It has a clear expected outcome, timeline, and action plan. It's the perfect way to organize tasks and assign resources for your operations plans. As you go forward, organize your top issues from steps 1-6 into initiatives that can be shared with the team.

If you're using the included workbook, it has this functionality built in and is a great place to start implementing change right away. It's intended to be altered and used as you see fit. Here's how it works.

Note: It helps if you have the workbook open as you read the next sections.

On the first sheet, enter the high-level information of each initiative that you'd like to create. You can have several cost drivers per initiative, but each should have its own line. The Workbook comes pre-loaded with examples.

After you've filled in the Initiative Setup, hit the "Run Initiatives" button to fill in the other sheets with all relevant information.

Prioritize Cost Drivers

In the next sheet, you'll see the initiatives and cost drivers. A few key points:

- The sheet carries over the data from the Initiative Setup sheet. The Baseline Metric is the value you have today, the Target KPI is what you want to be at. Enter the actual values over time as they change.

- A Risk Score is generated based on the Impact and Variance to target. A higher impact and a higher variance away from target will result in a higher risk score.

This sheet helps you quickly prioritize your cost drivers over time.

Delegate and Empower Teams

Discuss tasks and ask for leaders to take responsibility of specific areas using the RACI Matrix (Responsible, Accountable, Consulted, Informed). Tie tasks to your factory areas (e.g., ops lead handles internal processes from Step 5, sales team owns external extensions from Step 6 like go-to-market strategy). This reinforces ownership by having team members step up themselves instead of you directly telling them what to do. Feel free to change the categories as needed. For example, maybe you don't need an Ops Lead and want to add another category which is totally fine. This is just meant to be a template to help you get started and is fully intended for you to change to your needs.

Cultivate a Change-Ready Team. With ownership assigned via the RACI Matrix, spark a culture that thrives on change. Start simple with weekly meetings but let the team run the meeting as much as possible to continue the transfer of ownership. Kick off the meetings and then hand it over to team members to share wins and lessons learned. This builds buy-in fast, turning skeptics into allies without forcing heavy training or heavy guardrails that limit creativity and speed.

Remember the decentralized leadership strategy. Give your team members control over their own destiny, and you'll see faster results, more accountability, and a team that is eager to work hard and reap the rewards that come from success.

Note that the RACI Matrix template is pre-set with dropdowns for assignments and auto-calculates completion percentage with color highlights (green for high buy-in, red for gaps). The information from the Initiative Setup is copied over after hitting the Run button, making delegation effortless and tied directly to your mapped drivers.

Set Up KPIs and Monitoring Systems

Define 5-10 metrics for each of your Step 1 drivers, like inventory turnover or forecast accuracy, with baselines from Step 3 analysis and targets for 10-20% improvements. First use low-cost dashboards like Google Sheets or Excel (such as the included Workbook). If you want a more advanced option, try using RStudio and its dashboard abilities to create an adaptive live system that can further reduce the time needed for data manipulation.

The Workbook includes a KPI Dashboard template which auto-generates averages, trends (Up/Down arrows), standard deviation alerts for variability, and even a line chart comparing baselines to targets. This is meant to be a quick way to get you started, feel free to transform it to best fit your needs as you evolve.

Iterate and Adapt

Schedule monthly reviews to analyze variances using Step 3 tools like regressions, test pilots, and update your Step 4 process maps. Make it a living plan, use the Workbook's Iteration Roadmap template to timeline actions, track ROI variance (green positive, red negative), and cumulative savings for ongoing ROI visibility.

Over time, this process can create a culture of positive trial and error. This cycle builds momentum: Assess, adjust, advance, repeat. Small cycles compound big wins. Anchor it in your Factory Model: Cycle through Inputs (new data), Core Processes (optimization), Returns (savings), and Infrastructure to keep your strategy focused. Prioritize and shift resources around the factory so that all areas are being addressed.

You now have the data collection, analysis, and processes to fully support this culture of fast and flexible iteration. Over time, you'll build the infrastructure so that your team can focus on meaningful change with the results showing up automatically in the data and dashboards. Following this process will build a solid foundation that will have you ready to scale with technology. It's going to ensure that you're not in 95% of companies that struggle with new technology initiatives. Instead, you'll be leading the pack.

Although we're coming to the end of this book, there are many more resources available at https://www.jasonunlocked.com. I hope that you'll reach out if you have questions or need help in specific areas. I'm constantly adding more content across my website and social media accounts and would love to help more.

Before we close, we're going to run through a few more tips on how to get started in the next section.

Risks and Mitigations: Strategy stagnation? Encourage ownership, make the workbook available to everyone. Adapt it, iterate it, and encourage creativity. Team buy-in issues? Start with quick wins like running the Supplier Management tool for 5-10% savings in audits. Over-complexity? The Initiative Setup centralizes everything, so you're guided from the start. Make it your own, improve it as you learn more!

Conclusion and Next Steps

Conclusion and Next Steps

Congratulations, we've now journeyed through the proven 7-step framework to transform your business into a lean, optimized, and tech-empowered organization.

As you continue to work through your transformation, remember these common pitfalls to keep your implementation on track:

- Over-relying on new tech without testing basics and building a data foundation first can lead to those 95% failure rates we discussed.
- Always pilot small at first, like I did with sensors in brewing for quick iterations setting up bigger wins later.
- Ignoring team input during delegation risks resistance. Counter it by starting with those weekly habits we outlined in Step 7.
- Underestimating variability in data? Double-check with manual spot-audits early on.

Before we finish the book, I want to cover some additional topics that can help you take this process to the next level. First up is contingency planning.

Contingency Planning

Once you have a good feeling for this process, you can scale it out for better and more consistent results with contingency scenario planning. This is where you run several scenarios for each initiative, prediction, and projection. At a minimum, you want three contingency plans for each:

- Best Case
- Worst Case
- Nominal Case

Spend the most time on the nominal case (what you reasonably expect to happen most of the time). The nominal case is what we've focused on up to now. After you have solid plans in place, run sensitivity analysis and broaden out your assumptions. A sensitivity analysis involves changing one variable at a time to see how strongly that variable impacts your results. In this way, you can find the key variables that drive your results.

For example, increasing the number of suppliers may not move the needle much, but increasing the order quantity through strategic quarterly buying might drop costs significantly. Or, increasing finance terms to 60 days instead of 30 might not matter all that much, but reducing the past due customer list might have a significant impact on cash flow and business operations. Or, targeting younger buyers might not matter, but targeting buyers that are in warmer temperatures might be a game changer. This is what sensitivity analysis does, it broadens and deepens understanding of the key variables in your equations.

After the sensitivity analysis is done, go to the extremes. Look at how the model changes in the best case and the worst case to then get expectations of how good and bad things can be.

For example, using my copper analysis, if the trend of copper price continued higher by 10% over the next six months (best case), what does that do to our expected sales? If the price of copper drops 25% over the next six months (worst case), what would that do to the projections?

You can do this for all assumptions and variables but start with the variables that make the most impact first as identified by a sensitivity analysis. To go even further on contingency planning, define the variables and outcomes of your processes and prediction models into three categories:

- Known knowns.
- Unknown knowns.
- Unknown unknowns.

The known knowns category involves scenarios where you have data to back up all variables and predictions. These are typically processes completely in your control. For example, you're analyzing how a process change will impact the cost of a product being made. You should be able to get a close estimate of the outcome because you know the process and all variables and know the expected outcome (known known).

The unknown knowns category involves scenarios where you know the process, but you may not know all the information about the variables involved. For example, you may know that your sales process shows correlation to the price of copper, but you're not sure how the price of copper will move. (known unknowns). For the most important variables and processes, spend time trying to change the unknown into a known through research or small experiments such as tracking the price of copper over time and using analyst predictions for future movement.

The unknown unknowns is the hardest category and can often have the largest impact. These are the events that are extremely hard to foresee that can have a drastic impact on your business. For example, an important customer suddenly loses a contract and can't buy any more products. Or, a financial crisis arises, and your customer base can't access the money it needs to pay you. This category provides the ultimate risk analysis for your business to identify the key areas to protect and survive.

For example, at one point in my career I was responsible for a segment that had just 1 or 2 customers that were driving almost 90% of total sales. The business was based on a contract, so it was relatively secure. However, using this process it became clear that losing this contract was a massive risk, even though it was highly unlikely. We immediately began to deploy resources to diversify the business segment which helped grow the customer base and revenue. This was important because the economy ended up shifting and we did lose the contract, but because we had identified the risk and started diversifying, we were able to offset a good portion of the lost sales which kept us alive until we recovered the contract.

These contingency plans should feed directly into the business processes. For example, if you're implementing sensors to control a process make sure you have a contingency plan in place and make it known to the team. If critical sensors go down, the contingency plan should immediately start such as manually collecting build times in a spread sheet or on a piece of paper. At the end of the day, collect information from all builds and input into the system.

Another example, you shifted your data storage to the cloud which then went down. Switch your systems to use the local data backup storage until the cloud comes back online and connectivity is restored.

The main point is to avoid catastrophic outcomes by having a plan in place that your team can get up and running almost immediately. You're not trying to totally mitigate losses because building in total redundancy would take too much time and too many resources. The point is to stay alive until you can recover the main process or pivot to the next best option. You want to avoid the downtime and uncertainty that comes from catastrophic events; get your team working right away with solid contingency plans, sensitivity analysis, and known categorization.

Next Steps

First, if you didn't receive the included Workbook and Supplier Management Tool, head over to:

https://www.jasonunlocked.com/pages/bou-resources

If you haven't started already, here's how to get started with the next steps after this book:

- **Dive into the Workbook Immediately**: Use the included Lean Strategy Workbook templates, like the KPI Dashboard for tracking metrics or the RACI Matrix for team delegation, to apply Steps 1-7 to one high-impact process. Pilot it on a single cost driver from your factory model map for quick wins.

- **Begin the process**: If you haven't already, start going step by step through the process outlined in this book. Modify the Workbook as you go to best suit your needs. Look at the Supplier Management resource as an example of what can be done with some simple Macros and the help of AI.

- **Join My Community for Ongoing Support**: If you haven't already, subscribe to my newsletter at:

  https://www.jasonunlocked.com/pages/subscribe

- **Schedule a Review and Iterate**: Set a monthly check-in using Step 7's Iteration Roadmap to measure ROI and adapt to changes, involving your team early to build buy-in and culture.

- **Reach Out for Personalized Guidance**: If you want more personalized help, book a free 30-minute consulting call at:

  https://www.jasonunlocked.com/pages/consulting

- **Subscribe to my YouTube channel:** Another great resource to see hands on demos and walkthroughs of how I implement strategies and techniques like this, head to my YouTube channel and subscribe at:

  https://www.youtube.com/@JasonBUnlocked

As you embark on this path, remember the key themes of this book: start simple, become fast and flexible, seek knowledge of your business truths, decentralized leadership, a culture of experimentation. Test, learn, adapt, build your foundation, and then scale to propel your business to new heights.

Additional Reading: Beyond the Basics

You've now learned the 7-step arc, from mapping costs to synthesizing a lean strategy. To extend your gains responsibly and sustainably, consider these advanced topics.

AI Ethics: Responsible Implementation As you layer in AI (e.g., for regressions in Step 3 or automations in Step 5), prioritize ethics to build trust and avoid backlash.

Start with data privacy. Be careful about how you use and share data. Ensure your teams have a process and understanding of sensitive data, especially customer data. You are likely to have confidentiality agreements in place, make sure data is not disclosed unintentionally to the wrong people or publicly. Consider anonymized datasets when looking for patterns in your workbooks and comply with regulations like GDPR or CCPA. When making connections to third party services, ensure that your data stays private or find a way to anonymize it. This is not an exhaustive list by any means but is meant to get your mind working on what you may need to consider. As always, consult a professional for your specific needs.

Address bias by diversifying training data (e.g., check demographic inputs in logistic models from Step 6) to prevent skewed predictions favoring certain markets, customers, or suppliers. Double check to make sure the models don't have bias in your inputs or process that are causing the tools to overlook new directions because the data is too correlated to the incumbents. There are a lot of ways for biases to get in data sets, spend some time learning to ensure your tools are providing meaningful results.

Finally, promote transparency: Document AI decisions in your KPI dashboards, explaining how outputs like probability scores influence GTM strategies. AI moves quickly and allows us to get answers faster and we often skip this critical step. But, the more you document and share, the easier it will be to analyze and pivot later if needed.

Building this ethical approach can help reduce risks and create a team that's accountable, making tech a dependable partner in your work. It helps break down the opaque walls that can come up when using AI. When you're documenting the inputs and outputs it's easier to understand how the AI got to the conclusions it did. Then, you can analyze further to see if there are biases or incorrect assumptions. Never trust blindly, always verify as you go. And always consult a professional for your unique approach.

Sustainability: Green Optimizations Sustainability doesn't mean only doing what's right for the environment. Now, and going forward, sustainability will be a competitive advantage. The business that uses energy and resources efficiently will be the business that can charge a lower price and out-compete other companies.

This is why I focus on setting up the data collection and processes in steps 2-4 and the infrastructure in step 5. When you have the entire company aligned and ready for change, you can leverage these tactics for further gains.

For example, in Step 2's sensor deployments, opt for low-energy IoT devices to monitor resource use (like electricity), and target a reduction in electricity usage by 10-15%. This is what I did in my smart brewing project to target and optimize my process costs.

When you have the data, you can do things like match heavy electricity usage to low-cost grid times. For example, most utility companies will provide time periods where there is heavy use on their grid (typically 9 am to 5 pm) and some will offer discounted rates for utilizing the grid at non-heavy use times (such as overnight). You can use your data to align your processes such that they occur during the off-peak grid times and get immediate cost savings that also helps reduce the strain on the grid.

If possible, consider how your company might utilize power generation and storage on site. Again, it's not only about saving the planet. Companies that can generate their own power efficiently will enjoy cost savings and may even be paid by utility providers for extra energy generated. This drops right to the bottom line for the company.

As mentioned earlier, these topics and much more can be explored in much greater detail on my website and social channels. Finally, I'd like to say thank you for your support and thank you for reading and I hope to hear from you in the future.

Glossary

AI (Artificial Intelligence): Technology that mimics human intelligence for tasks like pattern recognition or automation, enabling tools such as prompts for data analysis (used throughout, e.g., in macro creation and process optimization).

Automation: The application of technology to perform repetitive tasks with minimal human input, such as using bots for data pulls (central to Step 2 and external integrations in Step 6).

CRM (Customer Relationship Management): Software systems for tracking and managing customer interactions, sales, and data (mentioned in data gathering and external partner management).

CSV (Comma-Separated Values): A simple file format for storing and exchanging tabular data between programs, ideal for importing exports from various systems (key in Step 2 for macros and translations).

Decentralized Leadership: Empowering teams at lower levels to make decisions, fostering agility and ownership while providing high level guidance and oversight (introduced in Step 4).

ERP (Enterprise Resource Planning): Integrated software for overseeing core business functions like inventory, finance, and supply chain (referenced in Step 6 for supplier and logistics coordination).

Factory Business Model: A framework viewing your business as a factory with inputs, core processes, outputs, and returns to identify cost drivers (core to Step 1).

Integrated Developer Environment: A software application that helps users write and test code. It often includes features such as highlighting key command words such as functions, check for proper syntax, autocompletion of code, compiling and debugging code, and more.

IoT (Internet of Things): Network of physical devices like sensors that collect and exchange data in real-time (used in Step 2 for monitoring).

KPIs (Key Performance Indicators): Measuring important outcomes like inventory turnover or sales growth. They are key because they are critical to achieving goals (referenced throughout, e.g., Step 7).

Macros: Automated scripts in tools like Excel to perform repetitive tasks, such as data cleaning (key in Step 2).

Pareto Analysis: A method based on the 80/20 rule to prioritize the vital few factors causing most issues or results (applied in Step 3 for insights).

Printed Circuit Board (PCB): A foundational component in electronics, consisting of a flat, insulating board (typically made of fiberglass or composite epoxy) with conductive pathways, usually thin copper traces, etched onto one or both sides. These pathways mechanically support and electrically connect electronic components (like resistors, capacitors, integrated circuits, and connectors) via soldering or other attachment methods. PCBs enable compact, dependable, and efficient circuitry, replacing older point-to-point wiring.

Programmable Logic Controller (PLC): A ruggedized, specialized digital computer designed for industrial automation and control. It serves as a solid-state control system that uses user-programmable memory to store instructions for executing specific functions, such as monitoring inputs (e.g., sensors detecting temperature or pressure) and controlling outputs (e.g., activating motors, valves, or lights) in real time.

RACI Matrix: A tool outlining who is Responsible, Accountable, Consulted, and Informed for each task or decision (included in the Lean Strategy Workbook for delegation).

Regressions: Statistical models for predicting patterns and finding correlations, like cost trends based on variables (applied in Step 3 for forecasting).

Value Stream Mapping: A visual technique to diagram the flow of materials and information, highlighting waste for lean improvements (recommended in Step 4 and Additional Reading).

Visual Basic: Visual Basic for Applications (VBA) is a user-friendly programming language integrated into Microsoft Office applications, including Excel, allowing non-experts to automate repetitive tasks without needing advanced coding skills. In Excel, VBA is commonly used to create macros that streamline workflows, such as automatically formatting reports, manipulating large datasets, or generating custom charts, helping business professionals boost efficiency and reduce manual errors.

Workflow: The sequenced steps in a process to complete work, often mapped and optimized with tools like flowcharts (optimized in Step 4).

Works Cited

Estrada, S. (2025, August 18). *MIT report: 95% of generative AI pilots at companies are failing.* Retrieved from Fortune: https://fortune.com/2025/08/18/mit-report-95-percent-generative-ai-pilots-at-companies-failing-cfo/

Gaus, T., & Schlotterbeck, M. (2025, May 01). *2025 Smart Manufacturing and Operations Survey: Navigating challenges to implementation.* Retrieved from Deloitte: https://www.deloitte.com/us/en/insights/industry/manufacturing-industrial-products/2025-smart-manufacturing-survey.html

Keen, E. (2025, June 17). *Gartner Announces the Top Data & Analytics Predictions.* Retrieved from Gartner: https://www.gartner.com/en/newsroom/press-releases/2025-06-17-gartner-announces-top-data-and-analytics-predictions

UiPath. (2025). *UiPath Helps MAS Holdings Save 14,000 Labor-Days Annually.* Retrieved from UiPath: https://www.uipath.com/resources/automation-case-studies/mas-holdings-manufacturing-rpa

About the Author:

Jason M. Brown is a seasoned engineer, entrepreneur, and executive with over 15 years of hands-on experience transforming operations and driving sustainable growth in the automotive, manufacturing, and mobility sectors. From designing life-saving helicopter armor to launching innovative high-voltage battery systems and scaling business segments from zero to multimillion-dollar revenues, Jason has mastered the art of turning complex challenges into profitable opportunities for companies big and small.

A Professional Engineer and MBA holder, he blends technical prowess—spanning Python, React, and AI integrations—with strategic leadership to optimize supply chains, boost efficiency, and foster innovative cultures. Whether through pro-bono consulting, pioneering a smart sustainable brewery in Michigan, or advising startups on market entry, Jason's philosophy is simple: Empower teams with data-driven tools to unlock their potential without the fluff.

When he's not decoding the next tech breakthrough, you'll find him exploring Michigan's craft beer scene or mentoring emerging leaders. Connect with Jason at jasonunlocked.com to explore how his proven frameworks can streamline your operations.